Manipulation of Exhaust Gas Values

Kai Borgeest

Manipulation of Exhaust Gas Values

Technical, Health, Legal and Political Background of the Emissions Scandal

Kai Borgeest
Hochschule Aschaffenburg
Zentrum für Kfz-Elektronik und
Verbrennungsmotoren
Aschaffenburg, Germany

ISBN 978-3-658-45863-8   ISBN 978-3-658-45864-5   (eBook)
https://doi.org/10.1007/978-3-658-45864-5

Translation from the German language edition: "Manipulation von Abgaswerten" by Kai Borgeest, © Springer Fachmedien Wiesbaden GmbH, ein Teil von Springer Nature 2023. Published by Springer Fachmedien Wiesbaden. All Rights Reserved.

This book is a translation of the original German edition "Manipulation von Abgaswerten" by Kai Borgeest, published by Springer Fachmedien Wiesbaden GmbH in 2017. The translation was done with the help of an artificial intelligence machine translation tool. A subsequent human revision was done primarily in terms of content, so that the book will read stylistically differently from a conventional translation. Springer Nature works continuously to further the development of tools for the production of books and on the related technologies to support the authors.

This Springer Vieweg imprint is published by the registered company Springer Fachmedien Wiesbaden GmbH, part of Springer Nature.
The registered company address is: Abraham-Lincoln-Str. 46, 65189 Wiesbaden, Germany

If disposing of this product, please recycle the paper.

Preface

September 19, 2015: The Tagesschau reports "VW faces billion-dollar fine". This report about the actions of American authorities against VW due to manipulated emission values in VW diesel vehicles was initially the tip of an iceberg, it not only had a long prehistory, but was followed by further scandals, also at other manufacturers. While initially the term "VW scandal" circulated through the media, now follow "emission scandal", "emission affair" or in the USA "Dieselgate".

Initially, I was only asked by people from my private environment and students what I thought about it; in essence, I frequently responded and also wrote in a letter to a newspaper that many car manufacturers would manipulate exhaust and consumption values in some way and that VW allowed itself to be caught by the US authorities. In 2015 and 2016, interviews followed on ZDF, ARD, and print media, which only allowed brief statements on the subject. In 2016, I was heard by the European Parliament committee of inquiry EMIS in Brussels, where I was able to take a somewhat more detailed position. This book is intended to provide even more detailed information. It probably wouldn't be difficult for me to write about 200 to 300 pages on this topic, as in previous technical books.

This little booklet is not intended to be a comprehensive textbook that presents all historical, health, technical, legal, and political aspects in detail; rather, it aims to present in a condensed form how the scandal came about, what it means for our health, how exhaust gas purification and engine control units work, what possibilities for deception exist, what is legal in Europe and the USA and what is not, which legal and political consequences follow, and how things can be improved in the future. The interested reader can then delve deeper into specific topics using the references provided in the book. Unlike my textbooks, I have preferred easily accessible sources, particularly on the internet (including the relevant legal norms of the EU), although at some points reference to scientific literature is sensible and necessary.

As an engineer and author of technical textbooks, who is accustomed to dealing with sober, non-interpretable facts, the presentation of political aspects represents a difficult but also appealing task for me. Of course, I also have my own opinion; however, by presenting different political viewpoints, I aim to help readers to form their own opinion. In favor of readability for the layperson, unlike in the textbooks, I have tried to use technical terms and formulas sparingly, and I ask the expert to forgive me for the sometimes necessary paraphrasing as a result.

I thank Mr. Felix Domke for providing Fig. 4.4, 4.5 and 4.6.

Aschaffenburg, Germany   Kai Borgeest
December 21, 2016

Preface to the Second Edition

Since the first edition of this book was published in 2017, much has changed. More unauthorized defeat devices have been uncovered. More realistic emission tests have been established. There are modern engines that comply with emission limits not only on the test bench but also in operation. The jurisprudence has evolved.

So it is time for a new edition of this book. The goal is to not let the volume grow too much despite the many new developments, and as in the first edition, to prioritize understandability for laypeople. Extensive literature references allow the reader to delve deeper into further technical, legal, economic, or political details.

Aschaffenburg, Germany
April 2021

Kai Borgeest

Preface to the Third Edition

After the first edition in 2016, I thought the second edition would be a review of the emissions scandal and a thematic conclusion.

Now the third edition is imminent, and there is no talk of a conclusion. Much has changed; with the new testing procedures RDE and WLTP introduced, it has become difficult to cheat on emissions. At the same time, the engines have evolved. Thus, many newly registered diesel vehicles are not only very low in emissions on the test bench, but also in traffic. This was achieved partly with many small improvements, but also with the new techniques presented in the book. At the same time, more and more locally emission-free electric vehicles have entered the market. So, despite the difficulties caused by the pandemic and the war in Europe, the engineers in the industry have now done their homework.

The legal processing in Germany is not going so well. As an expert in damage compensation proceedings, I have met many judges who have diligently familiarized themselves with the subject matter and wanted to clarify the disputed facts in their cases. However, I have also met individual judges and senates who allowed themselves to be misled by the lawyers of a vehicle manufacturer and with questionable party expert opinions in a transparent manner. The wave of proceedings that burden our courts hardly subsides. Because collective consumer protection in Germany is only weakly developed, the same facts are constantly being re-examined in thousands of parallel proceedings, without the courts having the necessary personnel capacities. The situation is even worse when it comes to criminal processing. Is the rule of law capitulating here?

Politically, little has changed so far. The control of political decision-makers by lobbyists, which made the emissions scandal possible in the first place, seems to be accepted by the participants. Transparency International recently found that there appears to be a lack of serious will to change this in Germany. A reduction in vehicle emissions is just a small part of a pollution and greenhouse gas reducing transport policy. A reorientation referred to as a mobility transition is not in sight, here we can only look enviously at many neighboring countries.

Thus, in this edition, there were minor improvements and updates in all chapters, with the section on now discovered defeat devices and the legal part having grown somewhat stronger.

If this book mentions series of engines and manufacturers, it would be a wrong conclusion to assume that these manufacturers have cheated more than others. However, there are manufacturers whose illegal defeat devices are now well known. Significant violation of the permissible limits have been demonstrated in the real driving operation of other manufacturers, and their control unit software may still reveal some interesting surprises.

Aschaffenburg, Germany   Kai Borgeest
February and June 2023

Contents

Glossary of Abbreviations

ADAC	Allgemeiner Deutscher Automobilclub e.V. (German Automobile Club)
AEUV	Treaty on the Functioning of the European Union
AGR	Abgasrückführung – Exhaust gas Recirculation
AO	Abgabenordnung (German Tax Regulation)
ARD	(German TV channel)
AUS	Aqueous Urea Solution
BGB	German Civil Code
BGH	Bundesgerichtshof (Federal Court of Justice)
BImSchG	Bundes-Immissionsschutzgesetz (German Immission Protection Act)
BImSchV	Regulation on the BImSchG
C	a higher programming language
CAA	Clean Air Act
CAN	Controller Area (a digital communication system)
CARB	Californian Air Resources Board
CF	Conformity Factor
CFR	Code of Federal Regulations
CJEU	Court of Justice of the European Union

CO	Carbon Monoxide
CO_2	Carbon Dioxide
CoP	Conformity of Production
CRT	Continuous Regeneration Trap
CVS	Constant Volume Sampling
DPF	Diesel Particulate Filter
DUH	Deutsche Umwelthilfe e. V. (German Environmental Aid)
EA189, EA288, EA896, EA897, …	Entwicklungsauftrag 189, 288, 896, 897, … (engine series affected by the emission scandal in the VW Group)
ECE	UN Economic Commission for Europe
EDC	Electronic Diesel Control
EG-FGV	EG-Fahrzeuggenehmigungsverordnung (European vehicle approval regulation)
EGR	Exhaust Gas Recirculation
EOBD	European OBD
EPA	Environmental Protection Agency
EU	European Union
EUDC	Extraurban Driving Cycle
EuG	Gericht der Europäischen Union (European General Court)
EuGH	Europäischer Gerichtshof (European Court of Justice)
e. V.	eingetragener Verein - registered Association
FSI	Fuel Stratified Injection
FTP75	Federal Test Procedure 75 (US Test Procedure)
HBEFA	Hand book Emission Factors for road transport
HC	Hydrocarbons
HCCI	Homogeneous Charge Compression Ignition
HFET	Highway Fuel Economy Test (US test for Highway Consumption)
ICCT	International Council on Clean Transportation—a research environmental organization

ISO	International Organization for Standardization—an international standardization organization
IUPR	In Use Performance Ratio
JC08	Japan Cycle 08
JRC	Joint Research Center
KBA	Federal Waterways Authority
LEV	Low Emission Vehicle
LG	Landgericht (German court category)
Lkw	Lastkraftwagen (Truck)
LNT	Lean NO_x Trap—NO_x Storage Catalyst
MIL	Malfunction Indicator Lamp
MNEFZ	modified NEFZ
NEDC	New European Driving Cycle
NEFZ	Neuer Europäischer Fahrzyklus (=NEDC)
NH_3	Ammonia
NMHC	Non Methane HC
NO	Nitric Oxide
NO_2	Nitrogen Dioxide
NO_x	Nitrogen oxides
N_2O	Nitrous Oxide
NTE	not to exceed
ÖAMTC	Österreichischer Automobil-, Motorrad- und Touringclub (Austrian Automobile, Motorcycle and Touring Club)
OBD	On Board Diagnosis
OLG	Oberlandesgericht (Higher Regional Court)
OME	PolyoxymethyleneDimethylEther
OPF	Otto Particulate Filter
OVC	Off Vehicle Charging—Charging from outside the vehicle (Plug-in)
PAK	polyzyklische aromatische Kohlenwasserstoffe (polycyclic aromatic hydrocarbons)
PAN	Peroxyacetyl Nitrate
PEMS	Portable Emission Measurement System
PM	Particulate Matter

RDE	**R**eal **D**riving **E**missions
rpm	**r**evolutions **per m**inute
SC	**S**ub-**C**ommittee (of an ISO TC)
SC03	**S**upplemental **C**ycle **03** (USA)
SCR	**S**elective **C**atalytic **R**eduction
SFTP	**S**upplemental **FTP**—Combination of FTP75, SC03, and US06
SiC	**S**ilicon **C**arbide
StGB	**St**raf**g**esetz**b**uch (German criminal code)
StPO	**St**raf**p**rozess**o**rdnung (German criminal procedure regulation)
TAFV	**R**egulation on **t**echnical **r**equirements for transport motor vehicles and their trailers
TC	**T**echnical **C**ommittee (of the ISO)
TCS	**T**ouring-**C**lub-**S**witzerland
TF	**T**ransfer **F**unction
THC	**T**otal **HC**
tkm	**t**housand **k**ilometers
TTIP	**T**ransatlantic **T**rade and **I**nvestment **P**artnership
UN	**U**nited **N**ations
USA	**U**nited **S**tates of **A**merica
US06	(American supplementary cycle)
VG	**V**erwaltungsgericht (German Administrative Court)
WHSC	**W**orld **H**armonized **S**tationary **C**ycle
WHTC	**W**orld **H**armonized **T**ransient **C**ycle
WLTC	**W**orldwide **H**armonized **L**ight-**D**uty Vehicles **T**est **C**ycle
WLTP	**W**orldwide **H**armonized **L**ight-**D**uty Vehicles **T**est **P**rocedure
WMTC	**W**orld **M**otorcycle **T**est **C**ycle
WWH	**W**orldwide **H**armonized OBD
ZDF	(German TV Channel)

Chronology 1

23.07.1973	The US Environmental Protection Agency (EPA) accuses VW of illegal defeat devices on carburetors of gasoline engines [EPA73]. These were physical defeat devices, not yet control unit software. In 1974, the EPA and VW agreed on a payment of 120,000 US$.
26.06.1991	The EU decides on a new European driving cycle (NEDC) for emission measurement [EU91/441], this was used until 2017, for consumption measurement partly longer.
30.11.1995	As a result of a settlement, General Motors must pay 11 million US$, because the control of the 4.9-l engines of the Cadillac brand enriched the mixture when the air conditioning, which was switched off during the test, was switched on. In addition, there are costs of at least 25 million US$ to remedy the problem and environmental compensation measures of up to 8.75 million US$ [USJ].
08.06.1998	Honda is supposed to pay a fine of 12.6 million US$ and 267 million US$ for remediation in the USA, because the detection of misfires was switched off [EPA98/08].
08.06.1998	Ford is supposed to pay 7.8 million US$ in the USA for an illegal defeat device. The application of the control unit caused excessive nitrogen oxide emissions during highway driving [EPA98/08].
22.10.1998	Caterpillar, Cummins, Detroit Diesel, Mack Trucks, Navistar International, Renault Trucks and Volvo Trucks must pay a total of 83.4 million US$ in the USA for defeat devices in trucks [EPA98/10].
23.06.2000	The magazine "Motorrad" first proves a defeat device on a motorcycle (BMW F 650 GS) [Schwarz00].
22.11.2005	Cabinet Merkel I: Federal Minister for Transport, Building and Urban Development becomes Wolfgang Tiefensee. Federal Minister for Environment, Nature Conservation and Nuclear Safety becomes Sigmar Gabriel.

K. Borgeest, *Manipulation of Exhaust Gas Values*,
https://doi.org/10.1007/978-3-658-45864-5_1

02.06.2008	According to a lawsuit in the USA [Cabraser16], Bosch warns VW that the shut-off function, which later triggers the emission scandal, is not permissible.
23.06.2008	For the first time, experiences with a mobile emission measurement system (PEMS) for passenger cars on a Fiat and a VW are published [Rubino08]. The validity of previous test bench tests is refuted.
28.10.2009	Cabinet Merkel II: Federal Minister for Transport, Building and Urban Development becomes Peter Ramsauer. Federal Minister for Environment, Nature Conservation and Nuclear Safety becomes Norbert Röttgen.
23.11.2010	The European Joint Research Center (JRC) presents the results of a measurement campaign in which many diesel cars emitted far more nitrogen oxides than the limits allowed. The Federal Minister of Transport did not react.
22.05.2012	Federal Minister for Environment, Nature Conservation and Nuclear Safety becomes Peter Altmeier.
17.12.2013	Cabinet Merkel III: Federal Minister for Transport and Digital Infrastructure becomes Alexander Dobrindt. Federal Minister for Environment, Nature Conservation and Nuclear Safety becomes Barbara Hendricks. Justice and Consumer Protection are combined into one ministry, Minister of Justice and for Consumer Protection becomes Heiko Maas.
03.11.2014	Hyundai has to pay a fine and compensation of 350 million US\$ in the USA due to manipulated consumption/CO_2 values [EPA14].
04.11.2014	The ICCT presents measurement results on real emissions from Euro-6 vehicles.
19.11.2014	Karl Falkenberg, Director-General for Environment of the EU Commission, reports the use of defeat devices to the industrial sector.
23.03.2015	The State Institute for Environment, Measurements and Nature Conservation Baden-Württemberg publishes PEMS measurement results that contradict the official emission values [LUBW15].
10.04.2015	With the words "I am at a distance to Winterkorn" the then chairman of the VW supervisory board Ferdinand Piëch criticized the CEO.
19.09.2015	**The news announces that VW in the USA faces a billion-dollar fine due to manipulations of emission values of the EA189 engine series.**
23.09.2015	The CEO of VW, Martin Winterkorn, resigns. He is succeeded by Matthias Müller.

15.12.2015	Frontal 21 shows in measurements at the Bern University of Applied Sciences with vehicles from various manufacturers that they exceed the emission limits by a multiple when driving a NEFZ outside the test stand.
17.12.2015	The European Parliament decides to set up an Inquiry Committee [EU16/34].
27.12.2015	Daniel Lange and Felix Domke present the "acoustic function" called defeat device from VW at the hacker congress 32C3 [Domke15].
22.04.2016	The KBA publishes a report on emission measurements of various models censored by car manufacturers [dpa16] [BMVI16]. Fiat and Chrysler show the largest exceedances of the limit.
12.05.2016	The news magazine Monitor reports on shutdown functions in the Opel Zafira. These were uncovered by Felix Domke.
14.06.2016	The Federal Ministry of Transport confirms to the member of parliament Scheuer (later himself Minister of Transport) that the KBA has stopped a necessary recall of Knaus Tabbert motorhomes at his instigation [Stoll20].
23.06.2016	The German Bundestag decides to establish an investigation committee [D18/8932].
07.07.2016	The investigation committee of the German Bundestag on the emission scandal constitutes itself.
02.08.2016	The Free State of Bavaria, as a shareholder, announces a lawsuit against VW [Hadem16]. The lawsuits were filed by Bavaria and Baden-Württemberg only later.
07.08.2016	The US authorities determine that emission-relevant functions of the 3-l engine of the VW Group are switched off after the duration of the American FTP75 test [EPA15].
01.09.2016	After a Fiat 500X exceeded the permissible limits by more than 20 times, the Federal Ministry of Transport intervenes for the first time and unsuccessfully informs the Italian Ministry of Transport (see 17.05.2017) that the exhaust gas recirculation is switched off after approx. 22 min [WiWo16/9].
19.09.2016	The organization Transport & Environment publishes a report on emission measurements on various Euro-6 models. Fiat and Suzuki show the largest exceedances of the limit, VW the smallest exceedances [T&E16].
08.12.2016	The EU Commission initiates infringement proceedings against 7 member states, including Germany. The reasons are the lack of sanctions and the concealment of relevant information by their authorities.

15.03.2017	According to a report by the French authority for competition, consumer and fraud control, Renault has been manipulating emission values since 1990 [Wittich17].
17.05.2017	The EU Commission opens an infringement procedure against Italy for non-compliance with approval regulations [EU17].
22.06.2017	The investigation committee of the Bundestag presents its final report [D18/12900].
25.08.2017	With James R. Liang, the first VW manager is convicted in the USA.
01.09.2017	For the type approval of passenger cars, an exhaust gas measurement in road traffic (RDE) becomes mandatory.
23.02.2018	Increased emission values were detected in BMW models. 11,700 vehicles are affected. According to company information, software for other models was accidentally used years earlier. [Bus18].
14.03.2018	Cabinet Merkel IV: Andreas Scheuer becomes Federal Minister for Transport and Digital Infrastructure, Svenja Schulze becomes Minister for Environment, Nature Conservation and Nuclear Safety.
13.04.2018	Herbert Diess (previously at BMW) replaces Matthias Müller as CEO of VW and remains in this position until August 2022.
17.05.2018	The EU Commission sues 6 member states, including Germany, for exceeding nitrogen oxide limits.
15.10.2018	The public prosecutor's office in Frankfurt am Main has premises of Opel in Rüsselsheim and Kaiserslautern searched. The subject matter is 95,000 diesel vehicles of the models Insignia, Zafira and Cascada, model years 2012, 2014 and 2017 with possibly illegally influenced software [MM18].
01.11.2018	The model declaratory action is introduced in Germany. On the same day, the Consumer Federal Association files a lawsuit against VW because of the manipulations on EA189 engines.
27.06.2019	Federal Minister of Justice and for Consumer Protection becomes Christine Lambrecht.
30.04.2020	After a settlement, the model declaratory action against VW is withdrawn.
02.07.2020	Daimler pushes a judge at the LG Stuttgart out of the diesel processes. His successor was previously employed by a law firm that also represents Daimler.
22.07.2022	The public prosecutor's office in Frankfurt am Main announces investigations at Fiat Chrysler in Germany, Italy and Switzerland. [Niesen20].

19.08.2020	An expert in a damage compensation process finds irregularities in a vehicle with a petrol engine [Schader20].
09.09.2020	The LG Braunschweig has admitted the indictment against Martin Winterkorn, Heinz-Jakob Neußer and other VW managers. [LTO20].
17.12.2020	The CJEU specifies the ban on shutdown devices and limits the argument of engine protection to cases where otherwise a sudden failure is imminent [EuGH20].
13.07.2021	The Hanover Public Prosecutor's Office once again searches the Hanover, Regensburg, and Schwalbach locations of the supplier Continental. [WiWo21]
08.12.2021	Cabinet Scholz: Federal Minister for Digital and Transport is Volker Wissing, Federal Minister for Environment, Nature Conservation, Nuclear Safety and Consumer Protection is Steffi Lemke. Federal Minister of Justice is Marco Buschmann.
27.04.2022	The Public Prosecutor's Office Frankfurt am Main has the business premises of Suzuki in Bensheim searched. Suzuki is suspected of having used illegal defeat devices in approximately 22,000 diesel vehicles (SX4 S-Cross, Swift and Vitara) with engines from Fiat-Chrysler (now Stellantis) [Gisevius22].
28.06.2022	The Public Prosecutor's Office Frankfurt on the Main has the business premises of Hyundai/Kia in Offenbach and Frankfurt searched. Illegal defeat devices are suspected in more than 210,000 vehicles of both brands [HR22].
08.11.2022	The ECJ decides that the German Environmental Aid (DUH) may sue against allegedly incorrect type approvals by the KBA [EuGH22].
21.03.2022	The ECJ has decided in case C-100/21 that owners of manipulated vehicles can claim damages without having to prove the manufacturer's intent.
20.02.2023	The DUH wins a lawsuit against the KBA for incorrect type approvals at the VG Schleswig [VG_SL].
16.05.2023	In the criminal proceedings against several Audi managers, the former Audi CEO Stadler also finally confesses, after a mild probationary sentence with a fine was offered to him.

The chronicle shows that manufacturers from all over the world have manipulated exhaust values and were convicted in the USA since there were limit values, VW had also previously been noticed. New is the extent of the fines in the billions and the number of vehicles affected. It is also noteworthy that insiders in business and politics were aware of the use of defeat devices before 2015.

References

[BMVI16]   Bundesministerium für Verkehr und digitale Infrastruktur: *Bericht der Unter-suchungskommission „Volkswagen"* (2016). https://bmdv.bund.de/blaetterk atalog/catalogs/235222/pdf/save/bk_1.pdf. Accessed: 12. Febr. 2023

[Bus18]   Business Insider: *Falsche Software bei 5er und 7er: Auch BMW könnte jetzt vor einem Abgasskandal stehen.* https://www.businessinsider.de/wirtschaft/ falsche-software-auch-bmw-koennte-vor-einem-abgasskandal-stehen-2018-2. Accessed: 20. Febr. 2023

[Cabraser16]   Cabraser, E.J.: Complaint. http://www.cand.uscourts.gov/filelibrary/1709/Con solidated_Consumer_Class_Action_Complaint.pdf. Accessed: 12. Febr. 2023

[Domke15]   Domke, F., Lange, D.: *The exhaust emissions scandal („Dieselgate")*, 32C3 (2015), Video: https://www.youtube.com/watch?v=d9HJw3AUvGk, Slides: https://fahrplan.events.ccc.de/congress/2015/Fahrplan/system/event_attach ments/attachments/000/002/812/original/32C3_-_Dieselgate_FINAL_slides. pdf. Accessed: 12. Febr. 2023

[dpa16]   Deutsche Presse-Agentur, published via heise online, *Abgas-Skandal: Enge Abstimmung von KBA und Herstellern* (11.11.2016). https://www.heise.de/ autos/artikel/Abgas-Skandal-Enge-Abstimmung-zwischen-KBA-und-Herste llern-3463860.html. Accessed: 12. Febr. 2023

[D18/8932]   Deutscher Bundestag: *Beschlussempfehlung und Bericht des Ausschusses für Wahlprüfung, Immunität und Geschäftsordnung (1. Ausschuss) zu dem Antrag der Abgeordneten ...*, Circular 18/8932. https://dip21.bundestag.de/dip21/btd/ 18/089/1808932.pdf. Accessed: 12. Febr. 2023

[D18/12900]   Deutscher Bundestag: *Beschlussempfehlung und Bericht des 5. Unter-suchungsausschusses gemäß Artikel 44 des Grundgesetzes*, Circular 18/ 12900. https://dip21.bundestag.de/dip21/btd/18/129/1812900.pdf. Accessed: 12. Febr. 2023

[EPA14]   EPA.   https://www.epa.gov/sites/production/files/2014-11/documents/hyu ndai-kia-cp.pdf. Accessed: 12. Febr. 2023

[EPA15]   EPA. https://www.epa.gov/sites/production/files/2015-11/documents/vw-nov-2015-11-02.pdf. Accessed: 12. Febr. 2023

[EPA73]   EPA, Environmental News, 23.07.1973. https://www.autosafety.org/wp-content/uploads/import/VWDefeatDeviceEPAProsecution7-23-73Pr.pdf. Accessed: 12. Febr. 2023

[EPA98/08]   EPA. https://www.epa.gov/sites/production/files/2014-06/documents/defeat. pdf. Accessed: 12. Febr. 2023

[EPA98/10]   EPA. https://www.epa.gov/enforcement/detroit-diesel-corporation-diesel-eng ine-settlement. Accessed: 12. Febr. 2023

[EU16/34] *Decision (EU) 2016/34 of the European Parlament from 17 December 2015 on setting up a Committee of Inquiry into emission measurements in the automotive sector, its powers, numerical strength and term of office.* https://eur-lex.europa.eu/legal-content/EN/TXT/PDF/?uri=CELEX:32016D0034. Accessed: 12. Febr. 2023

[EU17] European Commission: press release, Car emissions: Commission opens infringement procedure against Italy for breach of EU rules on car type approval. 17.05.2017 https://ec.europa.eu/commission/presscorner/detail/en/IP_17_1288. Accessed: 20. Febr. 2023

[EU91/441] *Counceil directive 91/441/EWG of 26 June 1991 amending Directive 70/220/on the approximation of the laws of the Member States relating to measures to be taken against air pollution against emissions from motor vehicles.* https://eur-lex.europa.eu/legal-content/EN/TXT/PDF/?uri=CELEX:31991L0441. Accessed: 12. Febr. 2023

[EuGH20] EuGH, Case C-693/18, Judgement of 17.12.2020

[EuGH22] EuGH, Case C-873/19, Judgement of 08.11.2022

[Gisevius22] F. M. Gisevius: *Suzuki im Abgasskandal unter Verdacht – Razzia der Staatsanwaltschaft Frankfurt,* https://www.anwalt.de/rechtstipps/suzuki-im-abgasskandal-unter-verdacht-razzia-der-staatsanwaltschaft-frankfurt-200187.html. Accessed: 12. Febr. 2023

[Hadem16] M. Hadem, H. Lossie: *Folgen des Dieselskandals: Bayern verklagt Volkswagen* (02.08.2016). https://www.autohaus.de/nachrichten/autohersteller/folgen-des-dieselskandals-bayern-verklagt-volkswagen-2728257. Accessed: 12. Febr. 2023

[HR22] Hessicher Rundfunk: *Verdacht auf Diesel-Betrug: Durchsuchungen bei Hyundai und Kia,* Hessenschau, 28.06.2022. https://www.hessenschau.de/wirtschaft/verdacht-auf-diesel-betrug-durchsuchungen-bei-hyundai-und-kia-in-offenbach-und-frankfurt,razzia-hyundai-kia-diesel-100.html. Accessed: 12. Febr. 2023

[LTO20] LTO: *Ex-VW-Chef Winterkorn muss vor Gericht,* 09.09.2020 https://www.lto.de/recht/kanzleien-unternehmen/k/lg-braunschweig-abgasaffaere-volkswagen-betrug-martin-winterkorn-anklage. Accessed: 1. Febr. 2023

[LUBW15] LUBW Landesanstalt für Umwelt, Messungen und Naturschutz Baden-Württemberg: *PEMS-Messungen an drei Euro 6-Diesel-Pkw, auf Streckenführungen in Stuttgart und München sowie auf Außerortsstrecken,* Bericht, Karlsruhe (2015). https://www.lfu.bayern.de/luft/luftreinhalteplanung_verkehr/projekte/euro_6_diesel/doc/pkw_euro6_abschlussbericht.pdf. Accessed: 12. Febr. 2023

[MM18] *Geschäftsräume in Rüsselsheim und Kaiserslautern durchsucht- Opel weist Betrugsverdacht nach Diesel-Razzia zurück.* Manager Magazin, 16.10.2018, https://www.manager-magazin.de/unternehmen/autoindustrie/opel-razzia-in-ruesselsheimer-bueros-wegen-abgasskandal-a-1233338.html. Accessed: 20. Febr. 2023

[Niesen20] Niesen, N.: Staatsanwaltschaft Frankfurt am Main, Press Release, 22.07.2020, *Durchsuchungen wegen des Verdachts des Betruges im Zusammenhang mit Diesel-Abschalteinrichtungen.* https://www.presseportal.de/download/doc

ument/5f18075c400000a29a7da782-sta22-07-2020.pdf. Accessed: 28. Jan. 2023

[Rubino08] Rubino, L., Bonnel, P., Hummel, R., Krasenbrink, A. et al.: *On-road Emissions and Fuel Economy of Light Duty Vehicles using PEMS: Chase-Testing Experiment*, SAE Int. J. Fuels Lubr. 1(1):1454–1468

[Schader20] Schader, N.: Manipulation auch bei Audi-Benziner?, Tagesschau (2020). https://www.tagesschau.de/investigativ/swr/audi-abgasskandal-109.html. Abruf 01.09.2020, nicht mehr verfügbar

[Schwarz00] Schwarz, W.: Topfreiniger. *Wie sauber ist die F 650 GS wirklich, fragte MOTORRAD in Heft 10. Jetzt musste die BMW samt Konkurrenz noch einmal auf den Abgasprüfstand*, Motorrad 14/2000, 23.06.2000. https://www. motorradonline.de/typen/technik-abgasreinigung-topfreiniger. Accessed: 12. Febr. 2023

[Stoll20] Stoll, R: *Minister Scheuer und Ex-Minister Dobrindt müssen im Fiat-Skandal mit Klage rechnen*, anwalt.de, 19.10.2020. https://www.anwalt.de/rechtstipps/ minister-scheuer-und-ex-minister-dobrindt-muessen-im-fiat-skandal-mit-klage-rechnen_181210.html. Accessed: 12. Febr. 2023

[T&E16] Transport & Environment: *Dieselgate 1ˢᵗ anniversary: all diesel car brands in Europe are even more polluting than Volkswagen – study*, 19.09.2016. https:// www.transportenvironment.org/press/dieselgate-1st-anniversary-all-diesel-car-brands-europe-are-even-more-polluting-volkswagen. Accessed: 12. Febr. 2023

[USJ] US department of justice: „U.S. ANNOUNCES $45 MILLION CLEAN AIR SETTLEMENT WITH GM FIRST JUDICIAL ENVIRONMENTAL RECALL – 470,000 CADILLACS", 95–596, 30.11.1995, https://www.jus tice.gov/archive/opa/pr/Pre_96/November95/596.txt.html.   Accessed: 12. Febr. 2023

[VG_SL] VG Schleswig, 3 A 113/18, Judgement of 20.02.2023

[Wittich17] Wittig, H.: *Renault Abgas-Skandal, Diesel-Betrug seit 25 Jahren?* auto motor sport, 16.03.2017, https://www.auto-motor-und-sport.de/news/renault-abgas-skandal-diesel-betrug-seit-25-jahren. Accessed: 20. Febr. 2023

[WiWo16/9] Schlesiger, C., Seiwert, M., Eisert, R., Wettach, S.: *Fiat 500X, Doblo und Jeep Renegade. Kommt der zweite Abgasskandal aus Italien?* Wirtschaftswoche, 06.09.2016.   https://www.wiwo.de/unternehmen/auto/fiat-500x-doblo-und-jeep-renegade-kommt-der-zweite-abgasskandal-aus-italien/14483066.html. Accessed: 12. Febr. 2023

[WiWo21] Wirtschaftswoche: VW-Dieselskandal: *Siebte Durchsuchung im Ermittlungskomplex Continental*, 14.07.2021. https://www.wiwo.de/unternehmen/ auto/jetzt-61-beschuldigte-vw-dieselskandal-siebte-durchsuchung-im-ermitt lungskomplex-continental-/27419818.html. Accessed: 1. Febr. 2023

Pollutants and their Effects

2

Since in the public often the terms *Climate protection* and *Environmental protection* are confused, it should be noted that the not noxious carbon dioxide (CO_2) probably contributes to global warming, so its reduction can contribute to climate protection; for the other exhaust gas components listed below, human health is the main concern, so their reduction cannot be called climate protection, but rather environmental protection.

Climate protection, the reduction of the greenhouse effect and CO_2 reduction are often mistakenly considered identical. The term greenhouse effect means that the earth's atmosphere, like a greenhouse, lets in more solar radiation than it lets out reflected radiation. This causes a temperature increase, which does not affect the whole world equally [NASA]. In addition to CO_2, other gases contribute to the greenhouse effect, e.g., the refrigerant R134a from air conditioning systems (no longer permitted in the EU, replaced e.g. by R744 or the controversial R1234yf), or the nitrous oxide N_2O (laughing gas) contained in traces in the exhaust gas. Besides the greenhouse effect, there are other effects that contribute to global warming, e.g., the change in backscattering (albedo) and the atmospheric heating by radiation-absorbing soot particles [Jacobson12].

With exception of pure hydrogen, today's fuels are liquid or gaseous hydrocarbons, possibly with a small proportion of bound oxygen (alcohols), additives and impurities. Ideal combustion of hydrocarbons oxidizes the carbon to carbon dioxide and the hydrogen to water (H_2O).

In reality the exhaust gas from combustion engines contains other pollutants, which are mainly formed as intermediates in incomplete combustion or by the oxidation of natural components of the air (nitrogen) and the fuel (sulfur). While the effects of individual pollutants are reasonably well known from experimental

© The Author(s), under exclusive license to Springer Fachmedien Wiesbaden GmbH, part of Springer Nature 2025
K. Borgeest, *Manipulation of Exhaust Gas Values*,
https://doi.org/10.1007/978-3-658-45864-5_2

(animal and cell culture experiments, simulations) and epidemiological studies (statistical observations on humans), little is known about possible interactions. These pollutants are primarily

- Particles,
- Carbon monoxide (CO),
- Nitric oxide (NO),
- Nitrogen dioxide (NO_2),
- Sulfur oxides (SO_x),
- Hydrocarbons,
- Aldehydes (R-COH, where R represents a hydrocarbon residue).

Particles are formed by incomplete combustion at low combustion temperatures or when, for example, due to late fuel injection, there is little time for combustion. In four-stroke engines, they consist of soot (carbon), on whose porous surface PAHs (polycyclic aromatic hydrocarbons) are deposited. These are hydrocarbons whose chemical structural formula consists of several benzene rings in the core, e.g., benzo[a]pyrene. While the carbon content of a soot particle is harmless, PAHs are highly carcinogenic. Homogeneously burning gasoline engines (intake manifold injection or carburetor) produce only small amounts of particles, gasoline engines with direct injection and diesel engines produce larger amounts without a particulate filter. The size distribution of the particles varies widely between a few nm and a few μm, a maximum is observable depending on the engine and operating condition between 50 and 100 nm, the legislation so far considers particles up to about[1] 2.5 μm/10 μm (shortly called PM2.5/PM10), from 2025 in the EU significantly smaller particles will probably be considered separately. The finer the particles are, the more difficult is the filtration and the easier they get into the lungs or from there into the blood. Epidemiological studies indicate an increased risk of heart attacks [Cesaroni14, Lelieveld20]. The immediate effect of ultrafine particles on the heart rate is the subject of current research [Rizza19].

Extreme quantities are produced by two-stroke engines with mixed lubrication (e.g., scooters, leaf blowers), although their particles have a different composition and are rather finer [Rijkeboer05]. Their composition is no less harmful because they contain high amounts of monocyclic aromatic hydrocarbons (in contrast to PAHs, only one benzene ring, e.g., benzene and toluene) [Platt14]. Road

---

[1] The legislation does not define a sharp limit, but a weighting function that decreases steeply around the maximum size [EU08/50].

traffic contributes significantly to particle pollution in Germany, however, since 2008, PM2.5 emissions from households and small consumers have overtaken those from road traffic in Germany, fortunately, this increase has not continued [UBA23P]. In households and small consumers, mainly heaters emit, among them solid fuel burners (wood, coal) emit significantly more than liquid burners (oil) and these more than gas burners [Ragland11]. In addition to exhaust particles, road traffic produces a considerable amount of abrasion particles, which are included in particle measurements. The abrasion contains hardly any PAHs, however, the smoke from wood burners, among other things, cannot be neglected in its danger compared to diesel exhaust due to its high PAH content [Danielsen11]. In addition to health effects, particles from diesel engines or wood burners also have a detrimental effect on climate change; how strong this influence is, especially in relation to CO_2, is still being researched.

Carbon monoxide (CO) reduces the blood's ability to transport oxygen and is produced by incomplete combustion due to a lack of oxygen. Carbon monoxide emissions have been well controlled since the introduction of exhaust catalysts, especially three-way catalysts for petrol engines (Sect. 4.4.1). Many environmental monitoring stations no longer record CO.

The **Nitrogen oxides** NO and NO_2 as well as traces of other oxides of nitrogen are produced at high combustion temperatures by oxidation of the nitrogen contained in the air to NO. They are mainly produced in diesel engines, to a lesser extent also in petrol engines with direct injection. Other engines are less affected. NO is a blood poison, it is short-lived and is oxidized mostly to NO_2 at not too high temperatures in the exhaust tract [Kolar90] or in the air. The characteristically pungent smelling NO_2 irritates the respiratory tract, contributes to acid rain and catalyzes the formation of atmospheric ozone (O_3), a strong irritant gas. Interestingly, remaining NO inhibits ozone formation, this can explain why the highest ozone concentrations do not occur near the exhaust release, but at some distance. A possible influence of nitric oxide on the development of cancer [Bonavida10, UBA18] and diabetes [UBA18] is being discussed. In the atmosphere, NO_2 can react with other air constituents to form peroxyacetyl nitrate (PAN). Apart from its greenhouse effect, its irritant effect and suspected other health effects, PAN is a long-term storage form of NO_2 that can transport NO_2 in bound form over long distances. In Germany, road traffic is the most important source, closely followed by energy supply [UBA23N]. A nitrogen oxide that is only released in traces, but is considered a greenhouse gas, is N_2O; it can, for example, be produced in exhaust aftertreatment at high temperatures, however, the proportion of road traffic in its formation remains low [Wiley].

Sulfur oxides irritate the respiratory tract and contribute to acid rain. Due to today's low-sulfur fuel, only very small amounts are produced, partly also by the co-combustion of highly sulfur-containing lubricating oil, which in a four-stroke engine, however, only occurs in small amounts, unless, for example, a defect in the cylinder head gasket or the turbocharger increases oil combustion. Emissions are high in two-stroke engines with mixed lubrication and in large engines that burn heavy oil. There are no exhaust emission limits for sulfur oxides, instead the sulfur content of fuels is legally regulated [EU09/30].

Hydrocarbons (short HC) only occur in small amounts in the form of unburned fuel residues in today's engines; with the exception of toxic and carcinogenic aromatics (benzene compounds), they are weakly or not harmful to health. Aromatics are still emitted in large quantities by two-stroke engines, these are described in [Platt14] as asymmetric polluters, as vehicles with two-stroke engines only represent a small part of the vehicle fleet, but contribute significantly to the air pollution with aromatics. Methane CH_4 is a strong greenhouse gas, but not harmful to health. Therefore, with Euro 5 for petrol engines, a separate limit value for NMHC (Non Methane Hydrocarbons, all hydrocarbons except methane) was introduced under the total limit value for hydrocarbons (THC, Total Hydrocarbons).

Aldehydes are oxygen-containing hydrocarbons that are produced by incomplete combustion, especially formaldehyde [Klingenb95]. They have different effects on health, but weaker than some other pollutants in the exhaust. Many aldehydes have a strong smell. There are no exhaust emission limits for aldehydes yet.

The **carbon dioxide** produced even in ideal combustion is not harmful to health, a suffocating effect is not to be expected as a component of engine exhaust gases; however, it is suspected that CO_2 contributes significantly to global warming through the greenhouse effect. The CO_2 emission is directly linked to fuel consumption, the combustion of one liter of diesel fuel produces 2.7 kg CO_2, the combustion of one liter of petrol 2.4 kg CO_2. A side effect of incorrect CO_2 determination is the loss of tax revenue in countries with a vehicle-related, CO_2-dependent taxation like Germany.

The list makes it clear that especially the typical diesel pollutants, namely nitrogen oxides and particles, are formed in opposite ways (nitrogen oxides at high combustion temperatures, particles with poor combustion) and thus pose a target conflict for the developer.

References

[Bonavida10]   Bonavida, B.: *Nitric Oxide and Cancer*. Springer, New York (2010). 2015 a similar book from the same author followed.

[Cesaroni14]   Cesaroni G. and 44 other authors: *Long term exposure to ambient air pollution and incidence of acute coronary events: prospective cohort study and meta-analysis in 11 European cohorts from the ESCAPE Project*, BMJ 2014; https://www.bmj.com/content/348/bmj.f7412. Accessed: 20. Febr. 2023

[Danielsen11]   Danielsen, P.H., Møller, P., Jensen, K.A., Sharma, A.K., Wallin, H., Bossi, R. and 6 other authors: *Oxidative Stress, DNA Damage, and Inflammation Induced by Ambient Air and Wood Smoke Particulate Matter in Human A549 and THP-1 Cell Lines*. Chem. Res. Toxicol. 24(2), 168–184 (2011)

[EU08/50]   *Directive 2008/50/EG of the European Parliament and the Council of 21 May 2008 on ambient air quality and cleaner air for Europe*. https://eur-lex.europa.eu/LexUriServ/LexUriServ.do?uri=OJ:L:2008:152:0001:0044:en:PDF. Accessed: 20. Febr. 2023

[EU09/30]   *Directive 2009/30/EG of the European Parliament end the council of 23 April 2009 amending Directive 98/70/EC as regards the specification of petrol, diesel and gas-oil and introducing a mechanism to monitor and reduce greenhouse gas emissions and amending Council Directive 1999/32/EC as regards the specification of fuel used by inland waterway vessels and repealing Directive 93/12/EEC*. https://eur-lex.europa.eu/legal-content/EN/TXT/PDF/?uri=CELEX:32009L0030. Accessed: 20. Febr. 2023

[Jacobson12]   Jacobson, M.Z.: *Air Pollution and Global Warming: History. Science, and Solutions*. Cambridge University Press, Cambridge (2012)

[Klingenb95]   Klingenberg, H.: *Automobil-Messtechnik*, Band C. Springer, Berlin, Heidelberg (1995)

[Kolar90]   Kolar, J.: *Stickstoffoxide und Luftreinhaltung*. Springer, Berlin, Heidelberg (1990)

[Lelieveld20]   Lelieveld, J., Pozzer, A., Pöschl, U., Fnais, M., Haines, A., Münzel, Th.: *Loss of life expectancy from air pollution compared to other risk factors: a worldwide perspective*. Cardiovascular research (2020) 116, S. 1910–1917. https://academic.oup.com/cardiovascres/article/116/11/1910/5770885. Accessed: 13. Febr. 2023

[NASA]   NASA: interactive map of climatic change. https://data.giss.nasa.gov/gistemp/maps. Accessed: 20. Febr. 2023

[Platt14]   Platt, S.M., El. Haddad, I., Pieber, S.M., Huang, R.-J., Zardini, A.A., Clairotte, M., Suarez-Bertoa, R., Barmet, P., Pfaffenberger, L., Wolf, R., Slowik, J.G., Fuller, S.J., Kalberer, M., Chirico, R., Dommen, J., Astorga, C., Zimmermann, R., Marchand, N., Hellebust S., Temime-Roussel, B., Baltensperger, U., Prévôt, A.S.H.: *Two-stroke scooters are a dominant source of air pollution in many cities*. Nature Communications 5, Article 3749 (2014). https://doi.org/10.1038/ncomms4749. Accessed: 20. Febr. 2023

[Ragland11]   Ragland, K.W., Bryden, K.M.: *Combustion Engineering*. CRC-Press, (2011)

[Rijkeboer05] Rijkeboer, R., Bremmers, D., Samaras, Z., Ntziachristos, L.: *Particulate matter regulation for two-stroke two wheelers: Necessity or haphazard legislation?* Atmos. Environ. **39**(13), 2483–2490 (2005)

[Rizza19] Rizza, V., Stabile, L., Vistocco, D., Russi, A., Pardi, S., Buonanno, G.: *Effects of the exposure to ultrafine particles on heart rate in a healthy population.* Science of The Total Environment. **650**(2), 2403–2410, https://doi.org/10.1016/j.scitotenv.2018.09.385

[UBA18] Umweltbundesamt: *Quantifizierung der NO2-bedingten Krankheitslast.* Final Report (2018). https://www.umweltbundesamt.de/sites/default/files/medien/421/publikationen/abschlussbericht_no2_krankheitslast_final_2018_03_05.pdf. Accessed: 13. Febr. 2022

[UBA23N] Umweltbundesamt: *Stickstoffoxid (NO$_x$, gerechnet als NO$_2$)-Emissionen nach Quellkategorien.* Chart (2023). https://www.umweltbundesamt.de/daten/luft/luftschadstoff-emissionen-in-deutschland/stickstoffoxid-emissionen#entwicklung-seit-1990. Accessed: 15. June 2023

[UBA23P] Umweltbundesamt: *Staub (PM2,5)-Emissionen nach Quellkategorien.* Chart (2023). https://www.umweltbundesamt.de/daten/luft/luftschadstoff-emissionen-in-deutschland/emission-von-feinstaub-der-partikelgroesse-pm25#emissionsentwicklung. Accessed: 15. June 2023

[Wiley] https://onlinelibrary.wiley.com/doi/full/10.1002/3527600418.mb1010244d0049. Accessed: 13. Febr. 2023

Testing Procedures 3

The limitation of emitted pollutants requires their realistic and comparable measurement. For this purpose, engine test benches and roller test benches with devices for measuring exhaust gas components are available. For a better understanding of the defeat devices, the former test procedures, which are relevant in many current civil and criminal proceedings, and of course the current test procedures are discussed here before the later chapter on law.

On the engine test bench, the engine is coupled to a load machine without the rest of the vehicle's drivetrain to simulate typical load profiles as they occur during driving. The load machine is a brake that is more precisely controllable and can be loaded for almost unlimited time, compared to, for example, a vehicle brake. Today, an electric machine (asynchronous machine) is commonly used as a load machine, which works as a generator and feeds the converted power back into the grid. It can also work as an electric motor and thus drag the combustion engine. This can simulate, for example, a downhill drive where the engine does not drive the vehicle, but brakes the downhill rolling vehicle. Engine test benches are precisely described in [Borgeest20M].

On a roller test bench, the entire vehicle with the complete drivetrain is tested, which is why it is also called a vehicle test bench. The wheels of the fixed vehicle stand on rotating rollers that simulate rolling on the road. The wheels are not steered. Vehicle test benches with a treadmill are rare. The rollers are coupled with load machines. The inertia of the vehicle during acceleration or deceleration is simulated by flywheels or electrically. Driving resistances due to incline or air resistance are also simulated by the load machines. On all-wheel test benches, all rollers rotate, while on most test benches only the drive wheels stand on a rotating roller, the other wheels stand still in contrast to a road drive. Although

K. Borgeest, *Manipulation of Exhaust Gas Values*,
https://doi.org/10.1007/978-3-658-45864-5_3

robots exist for driving test cycles, a trained human driver usually sits in the test vehicle, who is instructed by a monitor attached outside the windshield.

In the early development phase, engine test benches are used more often. Type approval (homologation) and the preparatory development in Europe takes place on engine test benches and on the road for trucks, on vehicle test benches and on the road for cars, and only on vehicle test benches for two-wheelers.

Engine test benches and vehicle test benches are equipped with similar or the same exhaust gas analyzers. The exhaust gas is taken from the tailpipe, diluted in a large pipe with a defined flow, and collected in bags (CVS, Constant Volume Sampling). For particle measurement, the exhaust gas is taken from the dilution channel without a bag. The contents of the bags are evaluated in measuring cabinets, each containing an analyzer for each pollutant to be measured. The technology is described in [Borgeest20M]. At research and development test benches, the samples are often analyzed directly without dilution and bags, allowing measurement values to be assigned more finely in time.

3.1 European Union

3.1.1 NEFZ

To get comparable results, a defined cycle, the NEFZ [EU91/441], also called NEDC, was driven for cars since Euro 1 (Fig. 3.1). It served for exhaust gas inspection and consumption measurement. The shift points are predetermined for manually shifted transmissions. The original warm-up phase of 40 seconds was omitted with Euro 3, for distinction the NEFZ without warm-up phase is sometimes called MNEFZ (modified NEFZ). It can be seen:

- The maximum speed reached only briefly is 120 km/h, although a large part of car traffic on motorways drives faster. This may be due to the fact that the permissible motorway maximum speed in almost all EU countries is in this order of magnitude, but it results in the vehicle operation with the highest nitrogen oxide emissions being disregarded. The high consumption in this area is also not taken into account.
- Accelerations from 0 to 50 km/h in 26 seconds are unrealistically low. By foregoing realistic acceleration, driving situations with an above-average emission of pollutants (especially particles) and also above-average consumption are excluded.

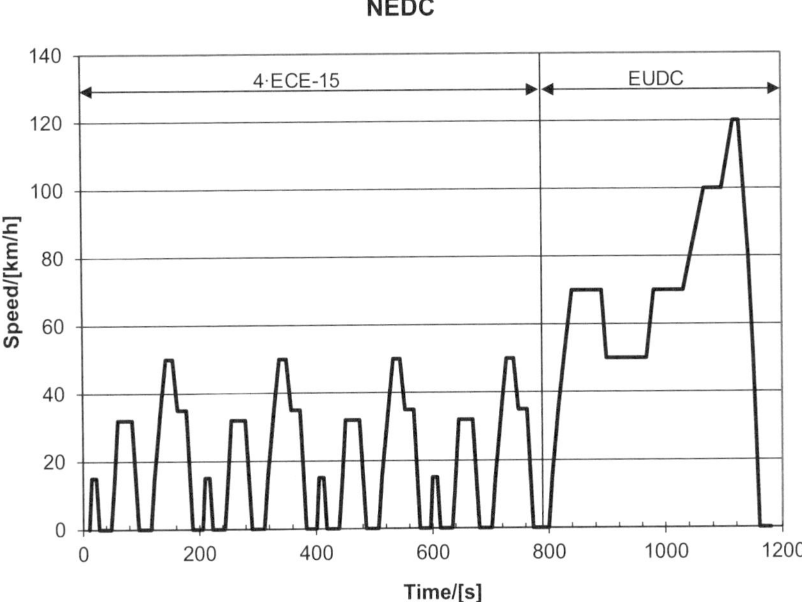

Fig. 3.1 Car test cycle NEDC (New European Driving Cycle), consisting of 4 city cycles (called ECE-15) and an extra-urban part (EUDC, extra urban driving cycle)

The NEFZ is driven at a temperature between 20 and 30 °C and an absolute humidity between 5.5 and 12.2 g of water per kg of dry air. The temperature is detected by the engine control unit via the intake air temperature sensor, there-fore some manufacturers have switched off their exhaust gas cleaning outside this temperature range, it then only works in a very limited temperature range (*temperature window* [BorgeestEU], see Sect. 4.5.1.5).

Sensors for humidity are only found in vehicles with air conditioning sys-tems, as they have higher tolerances than temperature sensors and are not directly connected to the engine control, they are less suitable for cycle detection; so far, no humidity windows comparable to the temperature windows for test stand detection have become known.

Before the NEDC, the test stand and the vehicle are prepared. A flywheel is installed on the test stand, which depends on the vehicle weight. [EU91/441] introduced eleven weight classes; some manufacturers managed to reduce the reference mass of the test object, mainly by removing parts, so that it falls into a lower weight class than in series condition. To adjust the driving resistance on the test stand, which is a vehicle-dependent function of speed, coast-down tests are carried out on a road beforehand ([Kadijk12] regarding consumption/CO_2, [Kadijk16] regarding pollutants). They are also called Road-Load-Tests according to their purpose [ISO10521]. Roads with a gradient (up to 1%, over a short section up to 1.5% permissible) are preferred for the coast-down test, as the vehicle rolls better there, so the apparent driving resistance is smaller. Higher roads in Southern Europe are particularly popular because the air pressure and temperature also influence the driving resistance. Since Euro 5, the results of this pre-test must be submitted for type approval, but they are not published. It would be interesting to investigate whether and to what extent possible coast-down modes in the software of vehicles have an influence. The unaccelerated driving resistance on the level is mainly caused by friction, especially on the tires, at low speeds, and mainly by air resistance, which increases quadratically with speed, at higher speeds [Borgeest20M]. Typical preparations on the vehicle were mounting low-loss tires with pressures up to 310 kPa (3.1 bar), minimizing the toe-in, possibly further adjustments to the chassis, pushing away the brake pads, adjustments to the transmission, filling in low-viscosity lubricating oils, running in the vehicle over several 1000 km, taping of gaps (air resistance) and removing the right outside mirror (air resistance).

The actual cycle delivers better results without auxiliary units (generator, air conditioning compressor), therefore the battery was fully charged beforehand.

The deficiencies of the NEDC led to proposals for more realistic cycles. A fundamental deficiency of any cycle is precisely its definiteness, which ensures repeatability and comparability, but on the other hand leads to the software of the engine control being able to recognize a cycle and thus switch the engine to a different behavior than in traffic. Furthermore, the many different driving situations, which lead to particularly high emissions in different operating situations for different vehicles, are difficult to integrate into a single cycle. Therefore, it makes sense to define additional maximum limits in addition to a cycle, which must not be exceeded in a real road trip (RDE, Real Driving Emissions).

An improved cycle, which included many real driving scenarios, was the Artemis cycle [André04]; it never made it into legislation.

3.1.2 WLTC

A further, more realistic cycle compared to the NEDC (without inclines) is the WLTC (Worldwide Harmonized Light-Duty Vehicles Test Cycle), defined in

[GTR15]. After its introduction was long delayed, it came into force in 2017. A conversion between NEDC and WLTC is difficult where old legal norms based on the NEDC are still in transition. The EU's CO_2MPAS program can be used for this [CO2MPASS]. The WLTC also feeds into the ECE regulation for emissions [ECE83] and for consumption and CO_2 [ECE101].

There are three variants of the WLTC, depending on the ratio of power in watts to vehicle mass in kg. European and American cars usually fall into class 3 (ratio ≥ 34), below that is class 2 and below that is class 1 (ratio ≤ 22). Despite its name, the WLTC is not valid worldwide, as a cycle that realistically represents traffic there has been established in the USA for many years. The cycles for class 2 and 1 (thin line in Fig. 3.2) are shortened compared to class 3, they omit fast and dynamic parts.

The WLTC is part of a test procedure called WLTP, which sets boundary conditions, e.g. temperatures. Important consumers are no longer disconnected

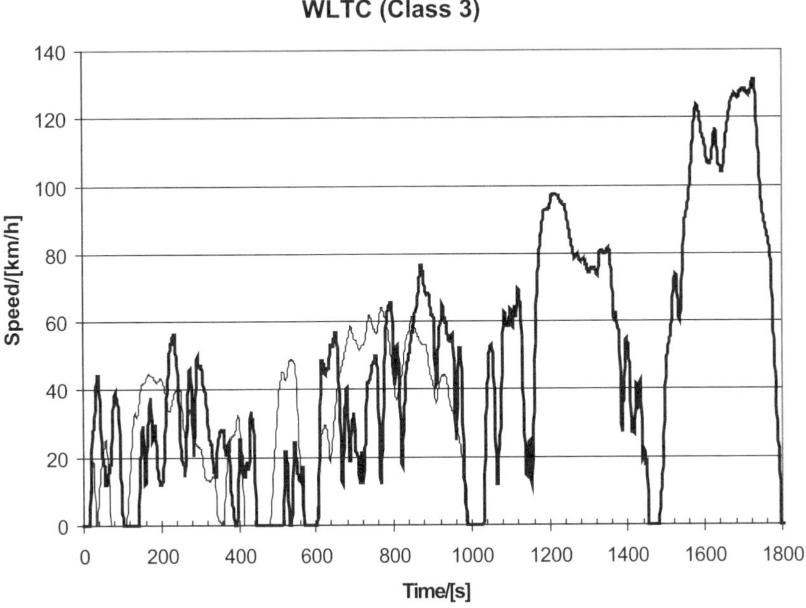

Fig. 3.2 Test cycle WLTC, class 3 (Worldwide Harmonized Light-Duty Vehicles Test Cycle). For comparison, the hardly relevant class 1 in Europe is *thinly* drawn in. The cycle differs slightly for vehicles with a top speed below 120 km/h

during the WLTP, the battery is 80% charged at the start of the test, the flywheels masses on the test stand are no longer diveded into step classes as in the NEDC. A realistic payload is taken into account in addition to the 100 kg weight of the occupants already included in the NEDC. The removal of parts to reduce the weight of the test vehicle as well as manipulations on the chassis or air resistance are no longer permitted. Manipulation possibilities in the preparatory road-load tests as in the NEDC are almost excluded; on the other hand, new uncertainties have arisen, as the road-load tests, which previously had to be carried out on roads, can now also be carried out under special laboratory conditions.

Details of the WLTP for plug-in hybrid vehicles are controversial. Controversial is a formula in Annex 8 of [ECE101], which applied in conjunction with the NEDC and continues to apply with the WLTP for the time being. According to this, the cycle is driven once with a full battery, once with an empty battery, the formula weights both fuel consumptions. It can lead to flattering values with a high electric range and ignores the energy supplied by the combustion engine or from the power grid to charge the battery.

3.1.3 RDE

The Cycle was supplemented for passenger cars from September 2017 for the reasons mentioned above by RDE [EU16/427, EU16/646, EU17/1151, EU17/1154, EU18/1832], i.e., emission measurements in a real drive of 90 to 120 minutes (real emissions have been measured in commercial vehicles for a longer time, see [EU16/1718]). Auxiliary consumers such as the air conditioning system are to be used realistically. Emission measurements in road traffic have only been possible since the exhaust gas measurement technology no longer required large measuring cabinets, but was miniaturized so that it can be carried along with the car to be tested (this has been possible with trucks for a longer time), such a device as in Fig. 3.3 is called PEMS (Portable Emission Measurement Systems). A special analyzer for hydrocarbons, the flame ionization detector [Borgeest20M], requires hydrogen for operation and is therefore unusual in car PEMS. Manipulations of the exhaust gas values by car manufacturers have been uncovered with PEMS.

The Euro-6-limits apply in the EU-RDE as binding maximum values (NTE values, not to exceed), which must not be exceeded during the entire "normal lifespan". Article (4) [EU07/715] assumes a normal lifespan of 100,000 km, but a maximum of five years, this assumption is likely to be exceeded frequently. The automotive industry was able to assert that the Euro-6 limits are softened by a "conformity factor" (CF). The CF is introduced in [EU16/427], its size for

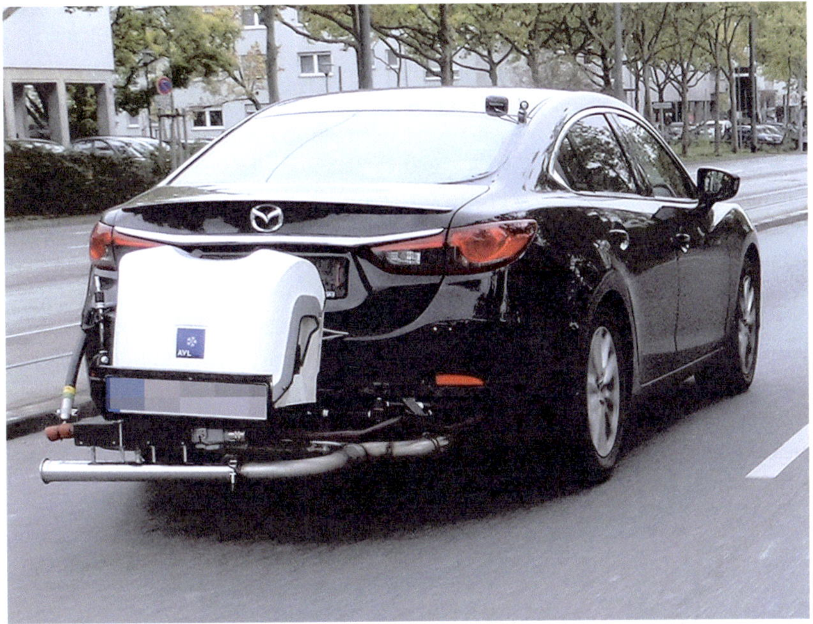

Fig. 3.3 Exhaust gas measurement with a PEMS. (Wikimedia Commons, User: LSDSL)

nitrogen oxides is defined in [EU16/646]. It is 2.1 from 2017 for type approvals, from 2019 for every new vehicle (Euro 6d-TEMP). This means that the NOx limit according to Euro 6 is increased from 80 to 168 mg/km in the RDE. From January 2020 (type approvals)/2021 (new registrations) the CF was reduced to 1.5 (Euro 6d). This value was further reduced to 1.43 with [EU18/1832]. A CF of 1.5 for the particle number was decided on 20.12.2016. Further adjustments were made in [EU19/1939]. The European General Court ruled on 13.12.2018 that the CFs are illegal [EuG18]. In Euro 6e from September 2023 (type approvals)/ 2024 (new registrations) this factor for nitrogen oxides is reduced to 1.1 and for particles to 1.34. This corresponds approximately to the possible measurement deviation compared to stationary exhaust gas measurement technology, which is why this factor is also renamed error tolerance.

There were also efforts to convert dynamic drives by a "transfer function" (TF); this would contradict the purpose of the RDE, to test realistically. The transfer function was abolished with [EU18/1832].

The RDE avoid a recognizable cycle, to which an engine could be trimmed; of course, the manufacturers will internally map the requirements set by RDE to a development cycle, which is significantly more demanding and no longer distinguishes between type approval test and real drive [Maschm16]. Manipulations in RDE tests are still possible by detecting the measurement setup on the vehicle. Even with the RDE, environmental conditions (altitude, temperature) are defined, but in a wider range than with previous test bench tests. In addition to the exhaust gases, the standard prescribes the measurement of other accompanying engine parameters, which are determined partly by additional sensors, partly also via the control unit with the help of existing sensors. Such external communication of the control unit could in principle be misused to detect the measurement by the control unit. There remains the risk of subsequent manipulations. The political development process of the RDE is justified and documented until 12/2016 in [EU16/WD].

3.1.4 Special Features of Hybrid Vehicles

Basically, hybrid vehicles, which combine a combustion engine with an electric drive, are tested in the same way as vehicles that only have a combustion engine. [ECE83] summarizes the special features in its Annex 14.

A crucial point is the state of charge of the drive battery at the start of the test. If the test were started with a full battery, in the extreme case the entire cycle could be driven electrically, the emissions would be 0. Therefore, hybrid vehicles are to be tested with a maximum charged and minimum charged battery. The results are to be weighted. The regulation distinguishes different cases, namely fixed or driver-influenced hybrid strategies as well as vehicles whose batteries can only be charged by the combustion engine or vehicles whose batteries can also be charged from the public power grid (OVC, **off** vehicle **c**harging, colloquially plug-in hybrids). Basically, all averages are structured according to the following scheme, whereby the exact meaning of the quantities in detail varies somewhat depending on the case considered:

$$m = \frac{D_1 m_1 + D_2 m_2}{D_1 + D_2} \tag{3.1}$$

m is the averaged mass of the pollutant considered, D_1 is the range with a full battery, D_2 is an assumed average range, e.g. 25 km, m_1 the pollutant mass with a charged battery (predominantly or purely electric operation) and m_2 the mass with a discharged battery (combustion operation). For a more precise meaning of the quantities in the respective individual case, reference is made to [ECE83].

In practice, it should be noted that hybrids without a power connection never and even plug-in hybrids rarely charge from the grid, but usually during the journey using the combustion engine. The formula can deliver unrealistically low emission values with a high electric range.

3.1.5 Cycles for Non-Car Engines

A test cycle for motorcycles similar to the WLTC is the WMTC [EU14/134].

For truck engines, different cycles are driven on the engine test bench, namely the WHSC and the WHTC [ECE49]. A test of entire trucks on the vehicle test bench is not requested by the legislator. The WHSC (World Harmonized Stationary Cycle) lets the engine run stationary for a longer time at selected operating points, the change is made by a ramp. The WHTC (World Harmonized Transient Cycle) is a transient cycle that includes a road trip with city, overland and highway components, with the highway component having low dynamics. A difficulty in defining truck cycles is the appropriate weighting of the route components; while a truck used for delivery participates in the more dynamic city traffic, a long-distance truck often travels a long distance at almost constant speed. A special feature of the truck cycles on the test bench is that instead of absolute, type-independent speeds, speeds and loads relative to the respective maximum value of the engine (also as a function of time) are specified.

For the emission test of combustion engines outside of road traffic ("Off Road"), legal sources of the EU and Japan base the standard [ISO8178-4], the US legislation partially follows it. It contains, in addition to the universal cycle B, various special cycles: C ("Off-road vehicles and industrial equipment"), D ("Constant speed"), E ("Marine applications"), F ("Rail Traction"), G ("Utility, lawn and garden"), H ("Snowmobile"). Manipulation would be possible here via the boundary conditions of the cycles. Many of these applications are hardly dynamic, the problem of the reality of cycles is less here than in road traffic. However, some applications can also be highly dynamic, think of a shunting locomotive, a snowmobile or special industrial applications.

3.2   Countries outside the EU

Switzerland has been applying EU rules since 1995 according to Sect. 3.1 in accordance with the relevant regulation there [TAFV]. In the United Kingdom, EU rules applied until 31.12.2020; it is becoming apparent that even after the British exit from the EU, its emission legislation will continue to be applied there.

In the USA has already been realistically tested since 1975 for American conditions with the FTP75 cycle (Federal Test Procedure 75), which is based on a simulated road trip in Los Angeles. It consists of three phases; a less dynamic cold start phase, followed by a more dynamic phase, finally after another warm start the same profile as after the cold start is repeated. The test would be unsuitable in Europe due to its low maximum speed of 91.2 km/h. For vehicles with air conditioning, an additional cycle (SC03) was introduced in 2007, in which the ambient temperature is raised to 35 °C. Also in 2007, another additional cycle, US06, was introduced, simulating an aggressive drive with a highway section up to 96 km/h. The pollutant emissions are calculated by weighting FTP75 (35%), SC03 (37%) and US06 (28%), this combination is referred to as SFTP (supplemental FTP). For vehicles without air conditioning, only FTP75 and US06 are weighted in the ratio 72/28. The consumption is weighted from the FTP75 and the Highway Fuel Economy Test (HFET) in the ratio 55/45, in addition, some pollutants are also limited in the HFET. In California, a more aggressive test (California Unified Cycle) is run instead of the FTP75. Details on test preparation and test execution can be found in the Code of Federal Regulations [CFR], especially in parts 1065 and 1066.

Also outside the USA and Canada, there are countries that use US cycles. Japan only replaced the older 10-15 mode test with the JC08 cycle in 2015. Since long distances are rarely driven by car in Japan, it represented the often congested city traffic. As early as 2018, this was replaced by the WLTC. Other countries that have introduced the WLTC include Australia and China [Dieselnet]. Russia has temporarily suspended emission standards.

References

[André04]   André, M.: *Real-world driving cycles for measuring cars pollutant emissions, Part A: The ARTEMIS European driving cycles.* Report INRETS LTE, Bd. 0411 (2004)

[Borgeest20M] Borgeest, K.: *Messtechnik und Prüfstände für Verbrennungsmotoren.* 2nd Ed., Springer-Vieweg, Wiesbaden (2020)

[BorgeestEU] Borgeest, K.: Statement in the Committee of inquiry EMIS of the European Parliament (16.06.2016). https://www.europarl.europa.eu/streaming/or. html?event=20160616-1500-committee-emis&language=or. Accessed: 13. Febr. 2023

[CO2MPASS] V. Arcidiacono, G. Fontaras, D. Komnos, J. Pavlovic, A. Tansini e. a.: CO2MPAS: *Vehicle simulator predicting NEDC CO2 emissions from WLTP.* https://co2mpas.readthedocs.io/en/stable/. Accessed: 13. Febr. 2023

[CFR] USA, Code of Federal Regulations, Title 40, Chapter I, Subchapter U. https://www.law.cornell.edu/cfr/text/40/chapter-I/subchapter-U. Accessed: 13. Febr. 2022

[Dieselnet] https://dieselnet.com/standards/. Accessed: 13. Febr. 2023

[ECE49] Regulation No 49 of the Economic Commission for Europe of the United Nations (UN/ECE) – *Uniform provisions concerning the measures to be taken against the emission of gaseous and particulate pollutants from compression ignition engines and positive ignition engines for use in vehicles.* Rev. 6 (2013). https://www.unece.org/fileadmin/DAM/trans/ main/wp29/wp29regs/2013/R049r6e.pdf + 7 Amendments (2013–2021). Accessed: 13. Febr. 2023

[ECE83] Regulation No 83 of the Economic Commission for Europe of the United Nations (UN/ECE) – *Uniform provisions concerning the approval of vehicles with regard to the emission of pollutants according to engine fuel requirements,* Rev. 5 (2015). https://www.unece.org/fileadmin/DAM/ trans/main/wp29/wp29regs/R083r5e.pdf + 14 Amendments (2016–2022). Accessed: 13. Febr. 2023

[ECE101] Regulation No 101 of the Economic Commission for Europe of the United Nations (UN/ECE) – *Uniform provisions concerning the approval of passenger cars powered by an internal combustion engine only, or powered by a hybrid electric power train with regard to the measurement of the emission of carbon dioxide and fuel consumption and/or the measurement of electric energy consumption and electric range, and of categories M_1 and N_1 vehicles powered by an electric power train only with regard to the measurement of electric energy consumption and electric range,* Rev. 3 (2013). https://www.unece.org/fileadmin/DAM/trans/ main/wp29/wp29regs/2015/R101r3e.pdf + 10 Amendments (2013–2022). Accessed: 13. Febr. 2023

[EuG18] EuG, Cases T-339/16, T-352/16 und T-391/16, judgement 18.12.2018

[EU07/715] *Regulation (EC) No. 715/2007 of the European Parliament and of the Council of 20 June 2007 on type approval of motor vehicles with respect to emissions from light passenger and commercial vehicles (Euro 5 and Euro 6) and on access to vehicle repair and maintenance information.* https://eur-lex.europa.eu/legal-content/EN/TXT/PDF/? uri=CELEX:32007R0715. Accessed: 13. Febr. 2022

[EU14/134] *Commission Delegated Regulation (EU) No 134/2014 of 16 December 2013 supplementing Regulation (EU) No 168/2013 of the European Parliament and of the Council with regard to environmental and propulsion unit performance requirements and amending Annex V thereof.* https://eur-lex.europa.

eu/legal-content/EN/TXT/PDF/?uri=CELEX:32014R0134. Accessed: 13. Febr. 2023

[EU16/427] *Commission Regulation (EU) 2016/427 of 10 March 2016 amending Regulation (EC) No 692/2008 as regards emissions from light passenger and commercial vehicles (Euro 6).* https://eur-lex.europa.eu/legal-content/EN/TXT/PDF/?uri=CELEX:32016R0427. Accessed: 13. Febr. 2023

[EU16/646] *Commission Regulation (EU) 2016/646 of 20 April 2016 amending Regulation (EC) No 692/2008 as regards emissions from light passenger and commercial vehicles (Euro 6).* https://eur-lex.europa.eu/legal-content/EN/TXT/PDF/?uri=CELEX:32016R0646. Accessed: 13. Febr. 2023

[EU16/1718] *Commission Regulation (EU) 2016/1718 of 20 September 2016 amending Regulation (EU) No 582/2011 with respect to emissions from heavy-duty vehicles as regards the provisions on testing by means of portable emission measurement systems (PEMS) and the procedure for the testing of the durability of replacement pollution control devices.* https://eur-lex.europa.eu/legal-content/EN/TXT/PDF/?uri=CELEX:32016R1718. Accessed: 13. Febr. 2023

[EU16/WD] European Parliament, *Working Document No 3 on the inquiry into emission measurements in the automotive sector – Chapter 3: Laboratory tests and real-world emissions* (16.12.2016). https://www.europarl.europa.eu/doceo/document/EMIS-DT-594073_EN.pdf?redirect. Accessed: 13. Febr. 2023

[EU17/1151] *Commission Regulation (EU) 2017/1151 of 1 June 2017 supplementing Regulation (EC) No 715/2007 of the European Parliament and of the Council on type-approval of motor vehicles with respect to emissions from light passenger and commercial vehicles (Euro 5 and Euro 6) and on access to vehicle repair and maintenance information, amending Directive 2007/46/EC of the European Parliament and of the Council, Commission Regulation (EC) No 692/2008 and Commission Regulation (EU) No 1230/2012 and repealing Commission Regulation (EC) No 692/2008.* https://eur-lex.europa.eu/legal-content/EN/TXT/PDF/?uri=CELEX:32017R1151. Accessed: 13. Febr. 2023

[EU17/1154] *Commission Regulation (EU) 2017/1154 of 7 June 2017 amending Regulation (EU) 2017/1151 supplementing Regulation (EC) No 715/2007 of the European Parliament and of the Council on type-approval of motor vehicles with respect to emissions from light passenger and commercial vehicles (Euro 5 and Euro 6) and on access to vehicle repair and maintenance information, amending Directive 2007/46/EC of the European Parliament and of the Council, Commission Regulation (EC) No 692/2008 and Commission Regulation (EU) No 1230/2012 and repealing Regulation (EC) No 692/2008 and Directive 2007/46/EC of the European Parliament and of the Council as regards real-driving emissions from light passenger and commercial vehicles (Euro 6).* https://eur-lex.europa.eu/legal-content/EN/TXT/PDF/?uri=CELEX:32017R1154. Accessed: 13. Febr. 2023

[EU18/1832] *Commission Regulation (EU) 2018/1832 of 5 November 2018 amending*

Directive 2007/46/EC of the European Parliament and of the Council, Commission Regulation (EC) No 692/2008 and Commission Regulation (EU) 2017/1151 for the purpose of improving the emission type approval tests and procedures for light passenger and commercial vehicles, including those for in-service conformity and real-driving emissions and introducing devices for monitoring the consumption of fuel and electric energy. https://eur-lex.europa.eu/legal-content/EN/TXT/PDF/?uri= CELEX:32018R1832. Accessed: 13. Apr. 2023

[EU19/1939] *Commisssion Regulation (EU) 2019/1939 of 7 November 2019 amending Regulation (EU) No 582/2011 as regards Auxiliary Emission Strategies (AES), access to vehicle OBD information and vehicle repair and maintenance information, measurement of emissions during cold engine start periods and use of portable emissions measurement systems (PEMS) to measure particle numbers, with respect to heavy duty vehicles.* https://eur-lex.europa.eu/legal-content/EN/TXT/PDF/? uri=CELEX:32019R1939. Accessed: 13. Febr. 2023

[EU91/441] *Council Directive 91/441/EWG of 26 June 1991 amending Directive 70/ 220/EEC on the approximation of the laws of the Member Statesrelating to measures to be taken against air pollution by emissions from motor vehicles.* https://eur-lex.europa.eu/legal-content/EN/TXT/PDF/?uri= CELEX:31991L0441. Accessed: 13. Febr. 2023

[GTR15] Global Technical Regulation No. 15 (*Worldwide harmonized Light vehicles Test Procedure*), 2014, Amendment 5 (2019). https://www.unece.org/filead min/DAM/trans/main/wp29/wp29r-1998agr-rules/GTR15-am5-For_Reg istry_EN.pdf. Accessed 13.02.2023

[ISO8178-4] TC70, SC8: *Reciprocating internal combustion engines – Exhaust emission measurement – Part 4: Steady-state test cycles for different engine applications,* 4th Ed., ISO 8178-4:2020

[ISO10521] TC22, SC34: *Road vehicles – Road load,* ISO 10521, Teile 1–2 (2006)

[Kadijk12] Kadijk, G., Ligterink, N.: *Road load determination of passenger cars.* TNO, Delft (2012). https://www.transportenvironment.org/sites/te/files/publications/Road load determination of passenger cars - TNO-060-DTM-2012-02014.pdf. Accessed: 13. Febr. 2023

[Kadijk16] Kadijk, G., Ligterink, N., van Mensch, P., Smokers, R.: *NO_x emissions of Euro 5 and Euro 6 diesel passenger cars – test results in the lab and on the road.* TNO, Delft (2016) https://repository.tno.nl/islandora/obj ect/uuid%3Ae34e9b34-94de-4bec-bf95-a2098f6b899d. Accessed: 13. Febr. 2023

[Maschm16] Maschmeyer, H., Beidl, C., Düser, T., Schick, B.: *RDE-Homologation – Herausforderungen, Lösungen und Chancen.* MTZ – Motortechnische Zeitschrift, 10/2016, pp. 84–91

[TAFV] *Verordnung über technische Anforderungen an Transportmotorwagen und deren Anhänger (TAFV 1) of 19 June 1995 (as at 1 February 2019).* https://www.fedlex.admin.ch/eli/cc/1995/4145_4145_4145/de. Accessed: 20. Febr. 2023

Emission Control, Control Units, and Defeat Devices

4

4.1 Control Units

Internal combustion engines in passenger cars and trucks and their exhaust systems contain many sensors today that capture physical quantities. Sensor signals are forwarded to an engine control unit (in diesel engines called Electronic Diesel Control, EDC), which controls actuators in the engine with this information so that the engine operates "optimally" (the problem of defining "optimal" will be discussed later). In V-engines with many cylinders, there can also be two engine control units (one control unit per cylinder bank). Sensors are, for example, speed sensors, temperature sensors for coolant, air and exhaust, injection pressure sensors, air mass sensors, lambda probes, knock sensors and NO_x sensors [Borgeest20E]. The control unit contains the electrical interfaces to the sensors, a digital computer on which all measurement, control, regulation and diagnostic functions (especially on-board diagnostics, Sect. 4.6) are implemented as software, as well as various auxiliary circuits, e.g. for the power supply of the computer. The engine control unit is often supported by additional control units, e.g. for exhaust aftertreatment, with close communication between the control units via digital data lines (e.g. CAN bus). Actuators in diesel or petrol engines are, for example, electrically controlled injection valves (injectors), valves for controlling the injection pressure, flaps in the air and exhaust path, glow plugs, spark plugs and dosing valves for urea solution.

Bosch calls its currently most widespread generation of diesel control units EDC17 (for petrol engines MED17) (see Fig. 4.1). An EDC17C46-2.7 is, for example, a special device from this generation for a specific 2-litre diesel engine from the EA-189 series from VW with a common rail injection system, which was initially at the centre of the emissions scandal. The previous generations

K. Borgeest, *Manipulation of Exhaust Gas Values*,
https://doi.org/10.1007/978-3-658-45864-5_4

EDC15 and EDC16 no longer play a role in the vehicles affected by the emissions scandal, nor does the subsequent control unit generation MDG1. Other important suppliers of engine control units besides Bosch (Germany) are Vitesco (Germany, the former powertrain division of Continental) with the EMS3, which is technically similar to the EDC17, and its predecessors such as the Simos PCR2 (also used in 1.6-litre engines of the EA189 series) as well as Denso (Japan) and Delphi (USA). Control units from Denso, a company whose history began in the Toyota Group and in which Bosch was involved for several decades, are mainly found in Japanese vehicles. A Japanese supplier of minor importance is Mitsubishi Electric. Delphi, once emerged from the General Motors Group, mainly supplied the US market, but the 1.2-litre engine of the EA189 series from VW also has a Delphi control unit. Since 2020, the former powertrain division of Delphi belongs to BorgWarner. Visteon, once emerged from the Ford Group, also mainly serves the US market. Before the merger with a Japanese company (2019), the supplier Magneti Marelli belonged to the Fiat Group, which also uses control units from Bosch. Magneti Marelli also supplies other vehicle manufacturers, including VW, to a small extent. The French supplier Valeo, which took over the control unit development from Johnson Controls in 2005, has little significance in this segment on the German market; one German customer is VW. The French company Sagem supplied numerous control units for French vehicles, but since the merger with Snecma in 2005 and the subsequent restructuring, the company is no longer active as an automotive supplier.

When a vehicle manufacturer like VW works with several suppliers, this complicates development, which has to work with a variety of different detailed solutions, but strengthens the purchasing department's position against the suppliers. The technically often inexperienced purchasing department has the most powerful role in the specification of the overall system and sometimes in individual, additional functions, then come the engineers of many car manufacturers and finally the supplier. So it is not surprising that often not a good, but the cheapest solution for the manufacturer is realized. This can, for example, be an illegal shutdown function instead of a mature exhaust gas purification.

A supplier like Bosch provides the hardware with a raw software, the behavior of which can be largely configured by adjustable data values up to complex maps. This configuration, called *calibration*, is mostly done by the vehicle manufacturer, in a few cases third parties are commissioned with parts of the calibration or also the supplier. This way, the vehicle manufacturer can adjust the functionality of a software within the framework determined by the supplier's variables, without the supplier having to release his code and without the vehicle manufacturer having to reveal to the supplier how he uses the software. In 2015, Bosch documented

Fig. 4.1 An EDC17 engine control unit

44 "sensitive functions" that offer "a special potential for non-compliant applications" [BR22]. However, the raw software can also contain functions that were specifically developed at the request of a manufacturer. If we look at the function for cycle detection shown in Fig. 4.4 as an example, the supplier provides a function here that sets a marker in the software when a time-dependent distance is undershot and a corresponding function that sets a marker when a time-dependent distance is exceeded. The vehicle manufacturer can then adjust both curves to the task, in this case around the test cycle. According to [Cabraser16], the acoustic function was calibrated by the supplier deviating from this usual procedure, but since this is rare and no advantages are recognizable for either side, this point of the complaint is difficult to understand.

The software of new control units has a modular architecture called AUTOSAR [Borgeest20E], which allows the manufacturer or third parties to develop software components without the participation of the supplier, as long as they do not directly access the hardware. Buyers can also use this opportunity to play their suppliers off against each other more strongly. Although the official history of this control unit architecture began in 2003 with the founding of a consortium (preliminary considerations are even older), control units without AUTOSAR were still produced in series for many years. By now, almost all control units brought into circulation have this architecture, but usually still with the classic division of labor between car manufacturer and supplier.

The optimal operation of the engine mentioned at the beginning is difficult to define. An engine should be economical, low in emissions, powerful, quiet or apparently loud in some vehicles [Welt17] and possibly meet further requirements, regardless of whether the vehicle is at the traffic light, participates in city traffic, drives over the country road or the highway. At the same time, the vehicle must be produced at the lowest possible cost. Due to these partly contradictory goals, the manufacturer has to set priorities, which can even be different for the same vehicle depending on the operating situation. Two questions arise, namely how legal requirements (emissions, noise) are met and which features make the vehicle sell better than comparable vehicles of the competition.

4.2   Injection and Combustion Processes

The internal combustion engine generates torque by burning fuel in the upper area of the cylinder (combustion chamber). The control unit can influence how much fuel and air are supplied to the combustion process and when, i.e., at which piston position, the combustion begins. This happens differently in diesel engines, direct injection and non-direct injection petrol engines.

In a non-direct injection petrol engine, a uniform mixture of fuel and air with a fixed composition but in a variable total quantity (depending on the accelerator pedal) enters the combustion chamber via the intake valve. The combustion of the mixture begins with the spark from the spark plug.

Direct injection petrol engines inject pure fuel near the spark plug (FSI, Fuel Stratified Injection, stratified charge) at partial load, the rest of the cylinder can be filled with air. Because not the entire cylinder needs to be filled with a fuel/air mixture, a small amount of fuel is sufficient; this explains the consumption advantage. The ignition timing must be precisely coordinated with the injection timing. A disadvantage is the short time between injection and combustion, which opposes complete mixing and therefore easily leads to incomplete combustion; in traffic, this manifests itself in increased particulate emissions from direct injectors. Due to the concentrated combustion, locally high peak temperatures occur, leading to the oxidation of the nitrogen contained in the air and thus also to increased nitrogen oxide emissions from direct injectors. The oxidation of atmospheric nitrogen ("thermal NO") is the only quantitatively relevant formation process in combustion engines, other mechanisms in the flame front, the "prompt NO" [Kolar90], as well as the combustion of nitrogen contained in the fuel play practically no role.

Diesel engines, which also work with direct injection today, also have a poorer mixing of air and fuel, which is why diesel engines also produce more particles. Due to the higher peak temperatures, diesel engines produce more nitrogen oxides. Due to the high compression of a diesel engine, temperatures of up to approx. 700 °C can already exist shortly before injection. Thus, the fuel ignites itself upon injection, an additional ignition as in the petrol engine is not necessary.

To reduce the amount of particles directly in the combustion process, air and fuel should be well mixed. A petrol direct injector can inject long before ignition to have enough time for mixing, but then the entire cylinder must be filled with mixture, the consumption advantage of direct injection is lost, near full load direct injection engines are operated in this way. New combustion processes that allow a similarly good mixing as in classic carburetor engines, e.g. HCCI (Homogeneous Charge Compression Ignition, [Basshuys17]), have been researched for years, the series introduction in diesel engines is not foreseeable, petrol engines with a related process are however recently on the market. Currently, the mixing is improved by constantly increasing the injection pressures (up to 3000 bar) in the direct injection petrol engine, but especially in the diesel engine, to reduce the size of the fuel droplets. These measures are supplemented by the generation of swirls in the combustion chamber.

The required amount of fuel, which is measured by the control unit, depends directly on the load, i.e., the torque demanded from the engine. Without interventions due to gear changes, the required torque depends on the gradient, frictional forces, air resistance, and the acceleration of the vehicle. The torque that the engine can deliver at maximum is dependent on the speed.

4.3 Exhaust Gas Recirculation

Decreasing the peak temperatures during the combustion results in less NO_x. For this purpose, a part of the exhaust gas is mixed with the fresh combustion air in diesel engines and many direct injection petrol engines, this process is called AGR (Abgasrückführung) or in English EGR (Exhaust Gas Recirculation). Two effects primarily influence the peak temperature:

- The fresh air contributing to combustion is diluted by oxygen-poor exhaust gas,

- Exhaust gas components, especially CO_2 and water vapor, absorb more heat at the same temperature than fresh air.

Although the exhaust gas is initially hotter than the fresh air at about 100 to 300 °C, the above two effects predominate and lead to a reduction in the peak temperature in the combustion chamber, which can reach up to 2500 °C in the diesel engine without EGR. The temperature of the recirculated exhaust gas is usually lowered by a cooler.

A common misconception among laypeople is that exhaust gas recirculation leads to the afterburning of unburned exhaust gas components. This is indeed the case to a small extent, but it is not the main effect of exhaust gas recirculation. In fact, the reduction in combustion temperature slightly increases the proportion of unburned substances in the exhaust gas.

Fig. 4.2 shows simplified two types of EGR. The high-pressure EGR, which can be regulated more quickly in most operating situations due to the shorter exhaust gas path, is a standard with which manufacturers have a lot of experience. The low-pressure EGR, which works with cooler and cleaned exhaust gas, is rarer, a combination of both recirculations even rarer.

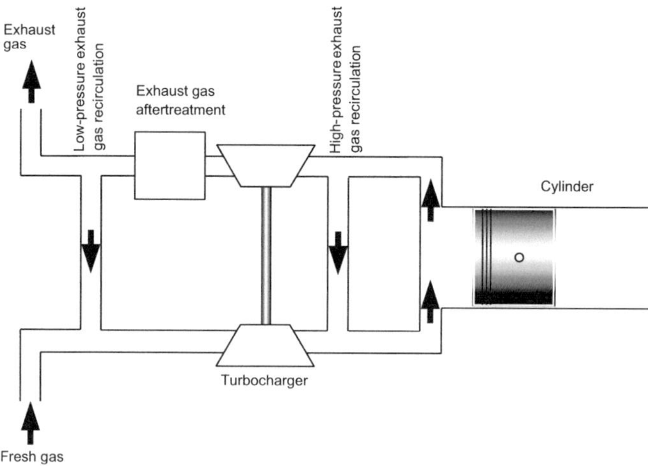

Fig. 4.2 Air and exhaust system of a combustion engine with turbocharger and exhaust gas recirculation

There, where the exhaust gas recirculation merges into the fresh air, a mixing device consisting of a fresh air and an EGR valve (exhaust gas recirculation actuator) is required (Fig. 4.3), which sets the ratio of fresh air and recirculated exhaust gas in the engine. There is also a cooler in the EGR. The low-pressure EGR usually also has a condensate separator and a filter.

In the exhaust gas recirculation, the principle "more helps more" does not apply, with increasing exhaust gas recirculation rate, a reduced performance is compensated with an increased fuel consumption. Not only do the momentarily occurring peak temperatures decrease - as intended - the temperatures also decrease over the remaining combustion duration, more particles are produced, which in turn burden the environment as well as the particulate filter and parts of the EGR. In particular, the EGR actuator easily gets sooty and can then jam; this is referred to as *sooting* or, depending on the consistency of the resulting layer, also as varnishing.

4.4   Exhaust Aftertreatment

In addition to in-engine measures, the exhaust gas can be subsequently cleaned of pollutants, in petrol engines by a three-way catalytic converter, in diesel engines and direct injection petrol engines by a particulate filter and a $DeNO_x$ filter for the reduction of nitrogen oxides and by oxidation catalyst. In diesel engines, particulate filters were introduced first, later also measures for the reduction of nitrogen oxides, especially with Euro 6. The exhaust aftertreatment is also controlled and regulated by electronics, either in the engine control unit or in a separate control unit that communicates with the engine control unit via the CAN bus [Borgeest20E].

4.4.1   Three-Way Catalytic Converter

In the 1970s, the three-way catalytic converter was introduced for petrol engines, despite initial resistance from industry, and has proven itself to this day. In this, three important chemical reactions take place simultaneously:

- Carbon monoxide is oxidized to carbon dioxide,
- remaining hydrocarbons are oxidized to carbon dioxide and water,
- Nitrogen oxides are reduced to nitrogen.

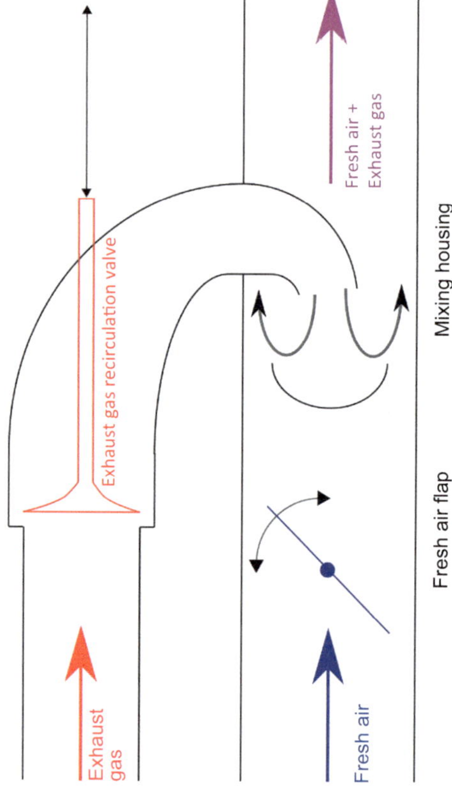

Fig. 4.3 Confluence of the exhaust gas recirculation into the intake path

A lack of oxygen makes the two oxidation reactions difficult, a high excess of oxygen makes the reduction difficult. If exactly as much air is available as is necessary for complete combustion of the fuel, the mixture is referred to as stoichiometric, in this case a 3-way catalytic converter works optimally. The quality of the mixture is described by the air-fuel equivalence ratio λ (lambda), it describes the amount of air in relation to the amount of air of a stoichiometric mixture. So it means

- $\lambda < 1$: lack of air (rich mixture),
- $\lambda = 1$: appropriate amount of air (stoichiometric mixture),
- $\lambda > 1$: excess of air (lean mixture).

To enable the optimal operation of the catalytic converter, the electronic λ-control of petrol engines was introduced quite early. An unregulated catalytic converter is spoken of when the engine does not yet have λ-control and a regulated catalytic converter when the engine (not the catalytic converter itself) has λ-control. If an engine without λ-control is operated rich, for example during cold start, the injection of additional air *(secondary air)* in front of the catalytic converter improves its effect. If an engine without λ-control is regularly operated lean, only oxidation can take place, so it can make sense to use an oxidation catalytic converter optimized for this reaction.

4.4.2 Oxidation Catalytic Converter

Oxidation catalytic converters only enable the oxidation of hydrocarbons and carbon monoxide, and to a small extent also of finest particles. They were partially used in earlier unregulated petrol engines. In diesel engines, they were used for a short transition period as the sole exhaust aftertreatment, today they are used in combination with $DeNO_x$ catalytic converters and with particulate filters. Often they are integrated with particulate filters in one housing to use the heat released during oxidation for heating the particulate filter to support regeneration. Oxidation catalytic converters, which mainly serve to heat subsequent components, are increasingly being supplemented or replaced by electric heaters.

4.4.3 Particle Filter

With Euro 5, cordierite particle filters have become established in the exhaust systemofdiesel engines. A particle filter acts as a fine-pored sieve that retains a portion of the solids contained in the exhaust stream. In the case of standard wall flow filters (where the exhaust stream is forced through a porous wall), this proportion is significantly higher than with retrofit filters. Particle filters are made of metal (usually porous sintered metal) or ceramic (cordierite or silicon carbide). They must be regularly (depending on the driving style after a few 100 km to just under 1000 km) burned free of their particle load, which consists mainly of soot; this process is called regeneration and should not take place during a measurement cycle because of the resulting pollutant peaks. Frequent regenerations with the associated peak temperatures (with over 1000 °C far above the exhaust temperatures) and temperature changes age the filter. In some particle filters, the regeneration temperature is lowered by ferrocene (a metal-organic iron compound) or cerium compounds. Since intimate contact of catalytic coatings with solid particles hardly takes place, these substances can also be added to the fuel as an additive. At sufficient temperatures and a sufficient supply of nitrogen oxides in the exhaust, the nitrogen oxides can also oxidize a portion of the soot particles, whereby they are reduced themselves, this effect has a quantitative significance especially in commercial vehicles; this effect can be supported by platinum or similar, catalytically active metals in the particle filter. A filter that has been optimized for the use of this effect is called CRT (Continuous Regeneration Trap) and is used in some commercial vehicles. If a particle filter can no longer be regenerated in the vehicle, there is still the possibility of maintaining the filter in the removed state by specialized companies, but the costs are over 100 € and the vehicle is not operational for the duration. Companies that make clogged particle filters passable (and ineffective) by drilling are acting legally under German law, but not the operators of the vehicles.

Since modern direct-injection gasoline engines emit significantly more particles than older gasoline engines, particle filters are also becoming established in gasoline engines (OPF, Otto Particulate Filter). These work similarly to diesel filters (DPF); an advantage in operation is that gasoline engines favor combustion due to higher exhaust temperatures, the low oxygen content in the exhaust gas of gasoline engines is a complicating factor.

4.4.4   DeNO$_x$ Catalyst

In addition to the NO$_x$-reducing effect of the 3-way catalyst in the gasoline engine (Sect. 4.4.1) and the CRT in some commercial vehicle diesel engines (Sect. 4.4.3), two types of special NO$_x$-reducing catalysts have become established in recent years, the storage catalyst in passenger cars, and the selective catalytic reduction (SCR) in trucks, buses and increasingly in more and more passenger cars. Since storage catalysts and SCR complement each other well with their different operating temperatures, both are also used in combination. Since all DeNO$_x$ catalysts have minimum operating temperatures above the ambient temperatures, no NO$_x$ reduction is effective immediately after a cold start. [EU08/692] required that the catalyst has a sufficient operating temperature 400 s after a cold start at -7 °C.

4.4.4.1 Storage Catalyst

A cost-effective solution is a storage catalyst (NO$_x$-Trap, Lean-NO$_x$-Trap or LNT), which stores nitrogen oxides by the contained ceramic (barium carbonate, barium oxide or some other metal oxides) reacting to barium nitrate. This reaction occurs from about 150 °C. Under very rich operating conditions (little combustion air), this reaction can be reversed for regeneration, this process takes place at intervals of a few minutes and lasts a few seconds. The nitrogen oxide released in the process is reduced by carbon monoxide, resulting in carbon dioxide and nitrogen, and residual carbon monoxide, which is not consumed in the reaction, can briefly escape. A detailed description of the storage reaction and releasing regeneration reaction for barium carbonate as well as the catalytically mediated reaction before storage (oxidation of NO to NO$_2$) is given by [Hauff13]. A storage catalyst is most effective immediately after regeneration. The problem is that sulfur oxides bind more strongly to the material than nitrogen oxides, but cannot be removed with the normal release reaction. Therefore, occasional special operating modes for desulfurization with rich mixture at exhaust temperatures above 650 °C are necessary, the safety distance to damaging temperatures is small, a temperature-controlled desulfurization therefore requires precise temperature measurement [Pott97]. A limited protection is possible by a preheated sulfur trap [Pott98]. During desulfurization, unpleasant smelling hydrogen sulfide can be released.

4.4.4.2 SCR Catalyst

Expensive, but with exception of cold start effective is the selective catalytic reduction. In this process, nitrogen oxides are reduced by ammonia (NH$_3$). This produces nitrogen and water. NH$_3$ is generated in the exhaust tract by blowing in or injecting an aqueous urea solution according to the standard [ISO22241] (internationally called AUS32, in Germany commonly branded as AdBlue). AdBlue costs about 1 €/l, depending on the packaging size the price can vary significantly up or down. Workshops charge up to 10 €/l including labor.

The solution is in a tank, which, although smaller than the fuel tank due to low consumption, is still considered a problem by car manufacturers due to its mass and volume. One variant in some older cars is filled by the workshop during service, another variant in most newer cars and in trucks and buses is filled by the driver himself and is recognizable by a blue cap. With unrestrictedly effective SCR, consumption varies greatly depending on driving style, from one to many liters of urea solution per 1000 km (since nitrogen oxide emissions are not constant and depend on the exhaust gas composition, different effective reactions between a third and a whole urea molecule per NO_x molecule in the SCR catalyst occur, an exact calculation is not possible). [Hüning19] estimates as a guideline 1% of fuel consumption. The actual consumption was noticeably lower in some vehicles. Also, a constant consumption independent of driving style is suspicious.

With increasing exhaust gas flow rate, which increases with engine speed, the dwell time in the catalyst and thus the conversion decreases; this can be compensated by a longer catalyst. It is questionable whether all installed catalysts are long enough to fully function even during brisk highway driving, which is not covered by test cycles.

Vehicle SCR catalysts work optimally between approx. 250 and 420 °C [Kolar90]. Catalysts for lower temperatures are not used in vehicles, as their optimal operating temperature is exceeded too often. To some extent, an SCR catalyst can store ammonia in this range to a limited extent, with increasing temperature the storage capacity decreases sharply, then SCR catalysts work with continuous reductant supply. Above 450 °C, NH_3 can be converted into additional nitrogen oxides, the exhaust gas cleaning then works against its actual purpose. Just below 1000 °C, the effectiveness of exhaust gas cleaning increases again, then a non-catalytic reaction also occurs. Above 1000 °C, the conversion almost comes to a standstill in vehicle catalysts. While the exhaust gas from diesel engines hardly reaches 1000 °C, temperatures above 450 °C are possible. Even higher temperatures can damage the SCR catalyst, this also depends on the choice of carrier material. An excess of ammonia can also lead to deposits of ammonium nitrate. To quickly reach the optimal operating temperature after starting, the SCR catalyst can be installed close to the engine instead of under the car floor. There are combinations of storage catalyst (for low temperatures) and SCR. Also combinations of an engine-near SCR catalyst and an SCR catalyst under the car floor are used, e.g. in VW engines of the EA288 Evo series (successor of the EA288 series). To not exceed the temperature limit, an additional extended pipe can be used for cooling, but this is not common for cost reasons.

4.5   Defeat Devices

In Chap.5 is discussed, under what conditions defeat devices are legal or illegal. A *defeat device* is no longer a physically tangible device today, but a part of the software in the engine control unit. It consists of a

- Detection of the test cycle or a road trip and
- Intervention in the engine or the exhaust aftertreatment.

The test cycle detection with corresponding vehicle influence is also called *Cycle Beating* in English. Not all defeat devices require a test stand detection (Sect. 4.5.2).

4.5.1   Cycle Detection

Some methods to determine whether the vehicle is running in the test cycle or participating in road traffic are:

- Steering (only on the road, not on the test stand),
- Driving profile (speed, load) deviates from test specification,
- Timing (clean operation only slightly longer than the duration of the test cycle after engine start under test conditions) or measurement of the corresponding driving distance,
- Recognition of cycle preparation,
- Temperature deviates from test specification (temperature window),
- Humidity deviates from test specification (application of this method not known so far),
- Air pressure deviates from test specification,
- Switching according to exhaust quantity,
- Age of vehicle or vehicle components,
- Engine hood contact (open during the cycle, closed on the road, no longer used today),

- Non-driven axle stands still on the test stand and rolls on the road (this detection is necessary because the anti-lock braking system would otherwise intervene due to seemingly blocking wheels, but it must not lead to interventions in emission-relevant assemblies),
- Navigation system does not detect movement despite turning wheels (application of this method not known so far).

These methods are often also combined.

4.5.1.1 Cycle Detection Via Steering Angle

During the exhaust gas measurement on the test stand, no steering takes place. A steering operation on the roller test stand can damage the test stand or the vehicle and is only possible in the standstill phases of a cycle. In road traffic, on the other hand, frequent steering is necessary. Thus, it can be detected by steering whether the vehicle is on the test stand or on the road. According to [Domke16], the VW Group has used a steering angle detection.

4.5.1.2 Cycle Detection Based on Driving Profiles

Especially via the driving profile hard-to-detect defeat devices can be realized, as it is normal and necessary for the engine control to adapt to the driving situation. The following examples show how this is possible. The first example makes the intention to obtain a vehicle's approval by manipulation particularly clear.

VW models with engines of the series EA189 with 1.2/1.6/2.0 l displacement were affected by the emission scandal besides other models. EA189 was the first engine series at VW with a "Common-Rail" injection system, which replaced the predecessor series EA188 still working with the older injection system "Unit Injector" from 2007, while other vehicle manufacturers had been using Common-Rail for a longer time. Under special conditions, the affected engines switch to an operating mode that intervenes in the exhaust gas recirculation and consumes particularly little AdBlue, VW has disguised this function with the name "acoustic function". In addition to some conditions with connected diagnostic tester, the following switching conditions lead into the clean test stand mode [Domke15]:

- Engine and fuel temperature above 15 °C,
- Atmospheric pressure exceeding 920 hPa, corresponding to a height above sea level of a maximum of about 750 m depending on weather conditions and
- Driving profile within certain limits according to Fig. 4.4.

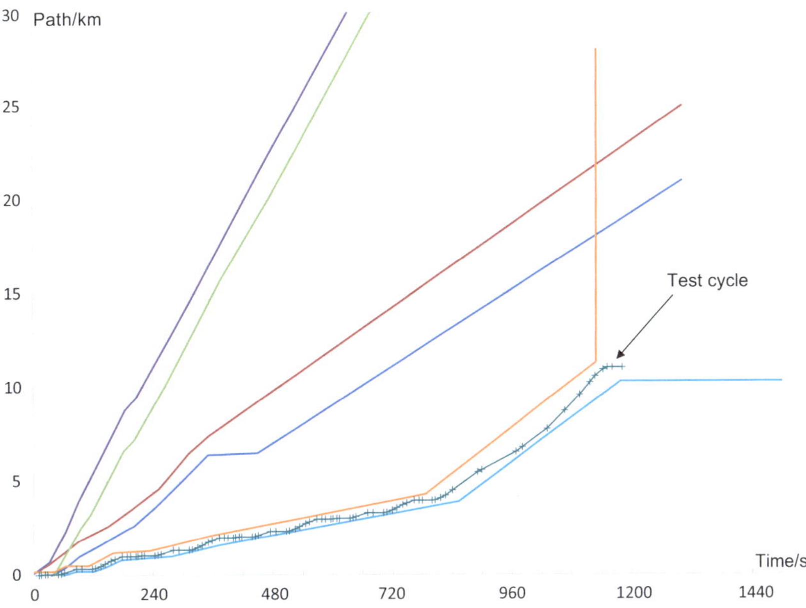

Fig. 4.4 Driving profiles of the VW "acoustic function". [Domke15]

These driving profiles represent the distance the vehicle must have traveled in a certain time after the start. Fig. 3.1 shows the NEDC test cycle in the usual form of speed as a function of time. If you calculate the integral of this function, you get the NEDC in another form, namely the distance covered as a function of time. This course can be precisely scaled into the lower driving profile in Fig. 4.4. A recognition of the test cycle thus takes place directly via the covered distance. The two other corridors allow an adaptation to further test cycles [Domke16].

Another function, also uncovered by Felix Domke, was found in the Opel Zafira (1.6-l diesel/Euro 6) and Astra (1.6-l diesel/Euro 6) [Monitor16]. This example also shows how functions that are sensible or even necessary are dated in such a way that they can be misused for cycle detection. This calibration was proven at Opel, but it is prototypical for a practice that probably also leads to the proven deviations between real emissions and test bench values at other manufacturers. One might discuss here whether it is an illegal defeat device or a borderline exploitation of unclear legal norms. There was also a "clean" test bench mode and a "dirty" everyday mode. The software switched depending on the injection quantity in relation to the speed. The chaotically appearing curves in Fig. 4.5 show the speed/quantity progression during a test cycle and in Fig. 4.6 during a road trip. The injection quantity can be considered representative of the torque demanded by the engine. If the driver exceeds the shutdown line, the "dirty" mode is switched on, after which he must first fall below the switch-on line again to return to the "clean" mode. The shutdown line is not fixed, but is defined differently for the 1st to 3rd gear (a), the 4th gear (b) and the 5th/6th gear (c). The picture shows that the shutdown line is never exceeded in the test cycle, but this happens more often during road driving.

Opel has published a statement on this function [Opel16]. Such a function, as Opel writes, makes sense, as the EGR would otherwise take away the air during strong acceleration, but the curve for reactivating the exhaust gas cleaning, which is quite far from the shutdown curve, is questionable, as well as the impression gained from the driven curves in Fig. 4.5a and c that the shutdown curves are laid around the test cycle. It also remains open why Opel uses the driving speed (with a threshold close to the cycle's maximum speed) as a secondary criterion for the catalyst temperature and does not directly evaluate its temperature.

4.5.1.3 Example: Cycle Detection with Timers

A very simple function was used by Fiat in the 500X model [DUH16]. There, the exhaust gas recirculation was switched off after 1300 s, which is slightly longer than the cycle lasts. Fiat did not convincingly explain why this was necessary for engine protection as claimed. A timer was also used in V6 engines in models of the brands VW, Audi and Porsche [EPA15], and there are indications in other vehicles. Such defeat devices are not new [EPA98/08]. A distance counter which switches off e.g. after 26 km works in a similar way. Named after a bit of an internal control unit register, such a defeat device at Daimler is called "Bit 15" [MSE].

4.5.1.4 Detection of Cycle Preparation

Before the cycle with the measurement is carried out, the vehicle is prepared. In the case of the NEFZ, this includes the obligatory repeated driving of the short high-speed component at the end of the cycle (Preconditioning, short Precon). After preconditioning, the engine is first switched off so that it can assume the ambient temperature in the test laboratory. Once the prerequisites are met, the actual test begins.

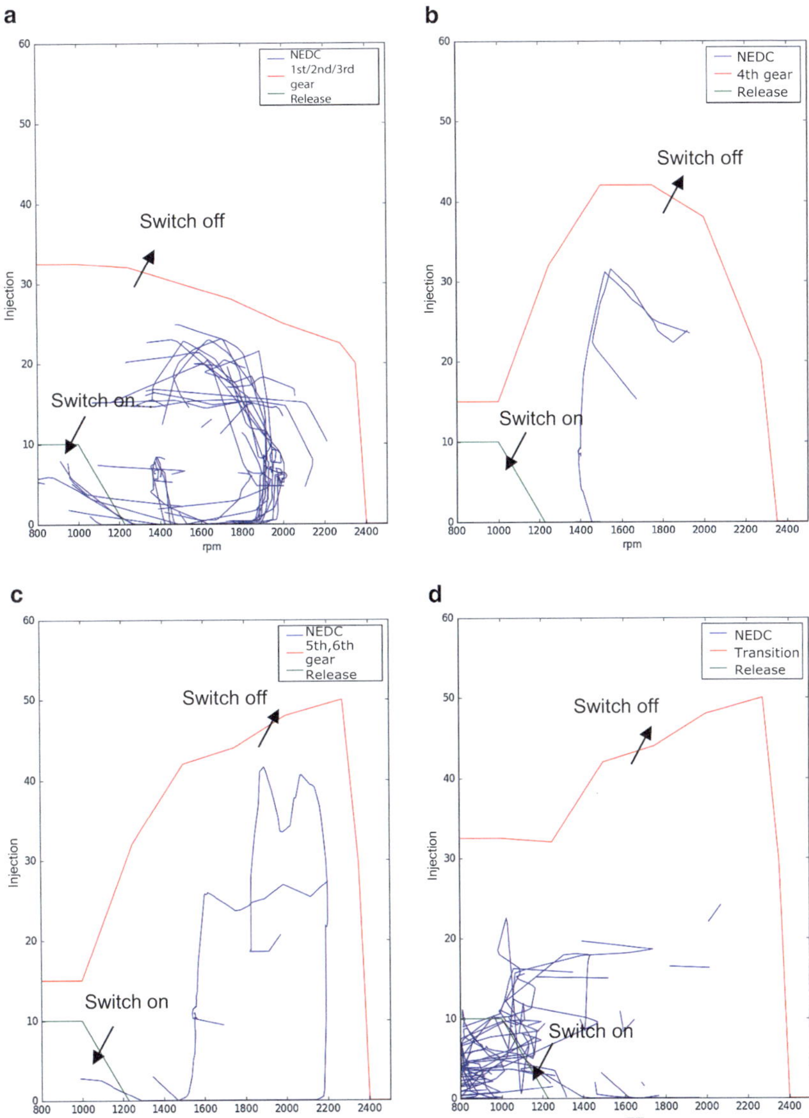

Fig. 4.5 Opel, switching off and reactivating the exhaust gas cleaning depending on injection quantity and speed (test cycle) at different gears (**a**, **b**, **c**) and in between (**d**). (*Source*: Felix Domke)

Fig. 4.6 Opel, switching off and reactivating the exhaust gas cleaning depending on injection quantity and speed (road trip). (*Source*: Felix Domke)

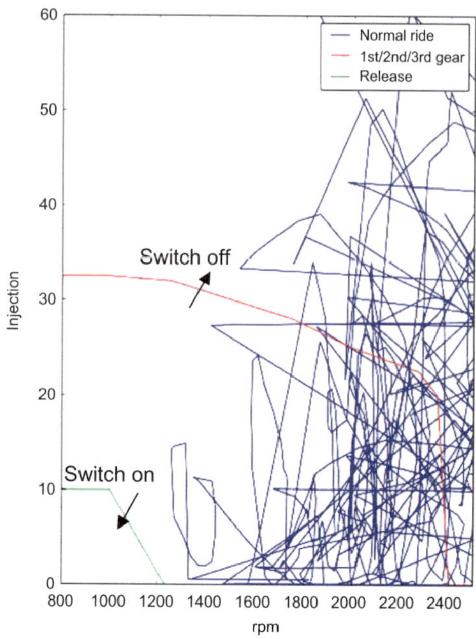

Since the Precon is a driving cycle itself (although not technically relevant), it can also be recognized as a cycle in addition to the actual driving cycle. This happens, for example, in vehicles with a storage catalyst, so that it can be thoroughly regenerated before the actual cycle.

4.5.1.5 Temperature Window

A very simple cycle detection, which has been and is used by many manufacturers, is a temperature window, also called a thermal window. In a temperature window, components of the exhaust gas purification are switched off when the temperature of the intake air and thus the ambient temperature is significantly below the lower temperature limit of the exhaust gas measurement (in the NEDC 20 °C) or significantly above the upper temperature of the exhaust gas measurement (in the NEDC 30 °C). In the WLTP, measurements are also taken at 14 °C, while road measurements have a larger temperature range.

Some manufacturers once adapted the temperature window very closely to the temperature range of the NEDC (e.g. with a 3 °C distance to the NEDC temperature limits), due to newer test procedures the temperature windows of newer models are often wider, but still present. Even after software updates due to the emission scandal, the temperature windows of older models often used by many manufacturers remained unchanged. Thus, there are manufacturers who define a normal temperature of +17 °C or lower in Central Europe as abnormal and switch off the exhaust gas purification for engine protection.

The temperature used for a temperature window is the temperature of the intake air or, if corresponding sensors are available, the outside temperature. At cold start, both temperatures are identical, with a warm engine the intake air is about 2 °C to 5 °C warmer. If the vehicle has been parked for several hours, as before a test stand run, the engine, the engine coolant and the lubricant also have approximately ambient temperature at cold start; this was used by Daimler in a defeat device [Domke20].

If the exhaust gas recirculation only operates in a narrow temperature range due to a temperature window, it can be produced more cheaply than if it were constantly used in road traffic. For example, it is cheaper to switch off the exhaust gas recirculation at low temperatures, rather than adapting it to the temperature during operation by means of a switchable bypass of the EGR cooler or other measures. At low temperatures, harmful condensates can form in the exhaust gas recirculation. According to common legal opinion, temperature windows are illegal [Führ16], also according to a ruling of the ECJ, *"software [..] which alters the level of vehicle emissions in relation to the driving conditions which it detects"* [EuGH20], which includes temperatures and thus temperature windows, is not permissible. In addition to the exhaust gas recirculation, measures for exhaust gas aftertreatment are also switched off with temperature windows. A temperature window is also illegal if it is not criticized by the KBA [VG_SL].

4.5.1.6 Air Pressure Window

Since most engine controls have sensors for the pressure of the ambient air, an air pressure window would also be possible analogous to the temperature window. The air pressure is used in some defeat devices, a defeat device that only uses the air pressure is unknown. Often, exhaust gas cleaning measures are reduced at low air pressures. This does not have to be illegal in every case. For example, the effect of exhaust gas recirculation is based on a dilution of the oxygen supplied to the engine. If you drive at high altitude in thin air, further dilution of the supplied oxygen by the exhaust gas recirculation would be counterproductive and could lead to smoking of the engine. However, the influence of the ambient air pressure on the operation of other measures such as SCR or storage catalytic converter would have to be questioned, as would a hard switch-over from a certain height represented by the air pressure.

4.5.1.7 Switching According to Exhaust Gas Flow

When the exhaust gas quantity increases, it flows faster through the unchanged chemically active surface of a catalyst. This reduces the time in which exhaust gas components can react, the conversion rate decreases. This means that in the SCR, fewer nitrogen oxides are reduced, while at the same time the ammonia slip increases. The catalyst must be large enough to work at any exhaust gas quantity, which depends on the speed. [Domke20] shows a function for a Daimler OM 642 series engine that specifies a lower target conversion rate at high exhaust gas flows, thus reducing the injection of urea solution. The same study shows a similar function that evaluates the nitrogen oxide mass flow.

4.5.1.8 Age of Vehicle or Vehicle Components

The age of a component can be indicated in km, operating hours or estimated wear. For example, the use of the SCR catalytic converter can be expressed as a percentage of its lifespan based on a simple simulation model in the control unit. The two functions mentioned in Sect. 4.5.1.7 reduce the quantity limits after a percentage aging. If the remaining life of the brand-new catalytic converter is given as 100%, [Domke20] shows that already at a remaining life of 99%, i.e. a still almost new catalytic converter, but not during the approval test, the two above-mentioned limit quantities are abruptly reduced and thus also the injection of the reducing agent.

4.5.2 Manipulative Interventions

If the control unit recognizes that the vehicle is not in a test cycle, one or more of the following interventions are carried out on conspicuous vehicles:

- Reduction of the EGR rate down to the shutdown of the EGR (Sect. 4.3),
- Reduction of the injected urea quantity for the SCR in the worst case down to 0 (Sect. 4.4),
- Heating strategy for the SCR,
- Regeneration of the storage catalyst on the test stand deviating from the driving operation (Sect. 4.4),
- The automatic transmission's shift program further exploits the engine's speed range,
- Coolant setpoint temperature control.

These measures increase the nitrogen oxide emissions outside of the test cycle.

By reducing the exhaust gas recirculation rate, the risk of sooting is reduced, and the required lifespan can be achieved more cheaply.

The reduction of the injected amount of urea solution allows for a smaller and lighter container for the urea solution and requires less frequent refilling. A special variant from Daimler, which does not work with a cycle detection, is presented by [Domke20]. It reduces the target conversion rate and the injected amount of urea solution, if a specified average consumption is exceeded.

The VW Group had problems with many of its six-cylinder engines (initially designated as EA896, then as EA897) to bring the SCR catalyst to operating temperature, especially after a cold start. Eventually, a heating strategy was developed that quickly generates the operating temperature on the test stand using additional fuel injections and oxidation catalysts. However, fuel is not injected in the same way for heating during driving operation, so this is also an illegal defeat device.

A regeneration of the storage catalyst deviating from the driving operation is intended to ensure that a test run starts with a freshly regenerated storage catalyst. This can happen during the preconditioning run. Since a regeneration does not last for the entire driving cycle and the course of the driving cycle is known, the manufacturer can set the regeneration immediately before the points in the cycle where the highest nitrogen oxide emission occurs. Characteristic of such a shutdown device is the seemingly random exceedance of the limits due to the initially random loading, when several similar (or also different) driving tests were carried out without the preparation. Such a pattern is recognizable in some engines of the EA288 series [BMVI16].

The dependence of the shift program on a road/cycle distinction is intended to improve the drivability of heavy vehicles at the expense of emissions. Audi has coupled a test stand detection with a modified shift program of automatic transmissions of the types AL551/AL951 [FAZ16, D18/13118]. The supplier ZF had to pay a fine of 4.2 million [Schwering20].

Daimler delays the warm-up of the engine in favor of the NO_x emissions on the test stand ("coolant setpoint temperature control") with engines of the OM 651 series [Traufetter21]. The central component is a heated coolant thermostat. By heating it during a detected test stand run, it reacts to a supposedly higher coolant temperature by a stronger temperature limitation.

All these interventions lead to increased emissions of nitrogen oxides compared to the values determined on the test stand. There are no known manipulative interventions that directly lead to significant differences between test stand and road in particle emission, but minor differences may be possible as a side effect.

However, sometimes considerable differences between test stand and road in fuel consumption and the resulting CO_2 emissions are known. To a large extent, the test conditions of the NEDC already allowed the manipulation of the measured values in a legal way during the test preparation (Section 3.1). Furthermore, manipulation is possible through a legally borderline design (similar to the example Sect. 4.5.1.2) of maps on the test.

4.6 On-Board Diagnostics

An electronic control unit like the engine control can monitor its own function, the function of connected sensors and actuators as well as system functions such as control loops, in which the control unit is involved. This diagnosis consists of two parts, the diagnosis relevant only for the service, which the manufacturer can design at his own discretion, and the diagnosis of emission-relevant errors, which is regulated by the European legislator (OBD, On-Board Diagnostics). The OBD was first regulated in the directive [EU98/69], which essentially supplemented an older directive on vehicle registration with a new annex. There have been some minor and major changes since then, a comprehensive revision took place with [EU08/692]. EU legislation now refers to [ECE83]. The manufacturer's own service diagnosis makes up the largest part of the diagnostic functionality in the control unit software [Borgeest20E], but the OBD is important for the topic of this book. Both parts of the diagnosis have in common that an error must first be recognized by the control unit, then the error must be stored in the control unit and it must be possible to read out the error with a diagnostic tester and to delete it from the error memory after its correction. Furthermore, it is useful for many errors to inform the driver about a detected error. In the case of manufacturer diagnostics, this typically happens via warning lights in the instrument cluster, but can also occur differently. With the OBD, it is prescribed that the driver must be warned about certain emission-relevant errors via a malfunction indicator lamp (MIL, yellow engine symbol in the instrument cluster), by flashing, continuous flashing, continuous lighting or acoustically. The control unit also registers for each emission-relevant error, how often the engine is in situations after clearing the error memory in which this error could be detected at all. Newer control units count in how many of all cycles since the last deletion by the OBD a fault was checked for (IUPR, In-Use-Performance-Ratio, exact definition see [EU11/582] with changes). This allows to check the validity of entered data, including whether the error memory was cleared shortly before reading out.

The OBD from the US legislator is called OBD 2, the one from the European legislator EOBD, both have a similar scope. A worldwide standardization (WWH-OBD according to [ISO27145], Worldwide Harmonized OBD) is planned and has been implemented for new commercial vehicles in the EU since 2014.

The diagnostic tester is connected via a standardized socket [ISO15031-3], which is usually in the driver's footwell. In the mandatory inspection in Germany, only the OBD was read out for several years and no emission measurement was carried out; since 2018, exhaust gases have also been measured at the tailpipe as before. Despite extensive regulation of the OBD, the manufacturer has leeway in deciding which errors are considered emission-relevant. Errors that lead to the exceeding of monitoring limits must be detected (Art. 4, [EU05/55]). Some errors, e.g. missing AdBlue, can lead to an intervention in the engine control up to a refusal of the engine start. The monitoring limits are much more generous than the type approval limits for a new vehicle, but they are approaching the type approval limits with newer standards [EU08/692, EU12/459, EU14/136, EU18/1832].

4.7   Possibilities for Improvement (Existing Vehicles)

A short-term solution for vehicles in which illegal defeat devices have been detected is a software update that deactivates the defeat device, thereby constantly putting the vehicle into the clean mode originally intended only for the test bench. The problem is that vehicle trials under realistic conditions were carried out in the dirty road mode with an effective defeat device, and therefore reliable information about lifespan and consumption after the update would only be possible after extensive retesting. While consumption after the update can be easily remeasured, a renewed lifespan test is unrealistically complex. The update process can be carried out within minutes, as the software with the associated data is stored in the control unit on a reprogrammable memory (flash) [Borgeest20E]; a similar process is carried out, for example, in the case of important software errors and is routine. Since the memory is protected against unauthorized write access, an update can only be installed by certain workshops with programming devices authorized by the manufacturer or by bypassing the protective measures (as some tuners do).

The 1.2-l- and 2-l-engines of the VW series EA189 received an update of the control unit software, in which the shutdown function is no longer active, the 1.6-l-engines additionally received a grid for stabilizing the air flow. VW states that the update, in addition to deactivating the shutdown function, optimizes the "injection characteristics", which includes changes to the injection, rail pressure, and charging. In particular, the fuel is injected at an increased pressure in the partial load range and an additional post-injection is introduced. The exhaust gas recirculation is also supposed to be regulated more precisely. According to the KBA, consumption does not increase, according to ADAC measurements on the Golf with the 2.0-l-TDI engine by 0.4% in the NEFZ and 2.5% outside, measurements by the ÖAMTC and TCS with an A4 showed no increase in consumption [Kroher16]. The risk of failure for components of the exhaust gas recirculation (especially actuators and coolers) and the particulate filter increases. After the update, owners report defects [WiWo16/10], in social networks, early failures of the EGR actuator are mainly reported; a causal connection is not known, but possible. The KBA certified VW that the lifespan of exhaust-relevant parts was not affected, but could not specify on request of the author how this was checked.

After Opel created media presence in May 2016 by denying, the company offered software updates for the Zafira and the Astra, later also for the Insignia [Consumer]. Largely unnoticed by the public, other manufacturers also carried out updates, either as part of an inspection or in connection with a recall.

A very effective retrofit solution is an SCR catalytic converter, but due to the necessary interventions in the engine control, this is difficult without the manufacturer's cooperation. Many manufacturers refused this cooperation, on the one hand, certainly to sell new vehicles, on the other hand due to unresolved questions about testing and liability in case of problems. Meanwhile, some manufacturers like Baumot, HJS, Dr. Pley SCR Technology GmbH and Oberland-Mangold offer retrofit solutions that do not require intervention in the engine control. A retrofit solution also requires space for an AdBlue tank, but even some vehicles, for which a retrofit would not be possible according to the vehicle manufacturer for this reason, have this space because they are equipped with SCR for export markets.

4.8   Possibilities for Improvement (New Developments)

Potential improvements include, among others, improved EGR control, improved sensors, switchable EGR cooling, low-pressure EGR or a combination of low-pressure and high-pressure EGR, filters and condensate separators, internal EGR, water injection (separate or emulsified with the fuel), non-sooting materials, burning off the EGR, concentrated, engine-close exhaust aftertreatment, variable compression and improvements in combustion processes.

The **precise tuning of the EGR** between nitrogen oxides and particles allows little leeway. The task of the control unit is to precisely regulate the optimal EGR rate (the proportion of exhaust gas in the air admitted by the engine) for each operating state. This works well in stationary engine operation, but rapid load or speed changes are more difficult. Due to the lack of dynamics in the NEFZ, this has not been a problem so far, but a real emission reduction requires control methods that also regulate well in dynamic operation [Stein16]; due to the research and development effort and the need for a sufficiently powerful computer in the control unit, not all technically possible measures have been taken in this regard so far.

A precise control also requires being able to **precisely measure the EGR rate**. So far, it has only been common to measure the sucked in fresh air mass using a hot wire sensor or a hot film sensor [Borgeest20E], the mass flow in the EGR is indirectly estimated by calculations from the control unit. Although mass flow sensors for the EGR have been available for many years (e.g. [Wienand06]), they are usually saved for cost reasons. Another problem is the lifespan of the air mass sensor, which can be detuned by dirt and oil vapors after just a few years. A detuned sensor would result in the wrong value of the EGR being precisely regulated, this can cause worse exhaust values than if the EGR were not regulated at all. However, suppliers are constantly working on improving their sensors in this regard.

A **switchable EGR cooling** helps to extend the usable temperature range of the EGR downwards. By cooling the returned exhaust gas, its density is increased and thus the mass of the returned exhaust gas in the cylinder. Cooling also reduces the initial temperature of the combustion process. Excessive cooling is undesirable, because then chemically aggressive condensate precipitates. If the returned exhaust gas already has a low temperature before the exhaust gas recirculation cooler due to the engine operating condition and the outside temperature, it makes sense to bypass the cooler through a valve to avoid excessive cooling. The necessary components (bypass valve, temperature sensor) increase system costs, in addition, a bypass valve is a component prone to failure, which can reduce system reliability. Suppliers now offer complete modules with exhaust cooler, bypass and bypass valve, which reduce the additional costs compared to individual solutions and are sufficiently reliable for practical use through appropriate testing.

Since the **low-pressure EGR** takes the exhaust gas after the aftertreatment, it is cleaner compared to the exhaust gas which ist taken directly from the outlet with a high pressure EGR. However, since condensation is likely at the lower temperatures of the low-pressure EGR, the water must be separated, otherwise the compressor of the turbocharger would be damaged by water droplets. The problem is successfully solved by the combined use of separators and filters. For cost reasons, this has been omitted so far in the high-pressure EGR and instead a temperature window has been defined, which in some vehicles switches off the EGR at normal outside temperatures.

The **internal EGR** is free of soot-prone components. It works without a connecting line between the exhaust path and the fresh air path. The recirculation takes place by keeping the exhaust valve open for a defined period of time during intake, this sucks a part of the exhaust gas from the prior combustion back into the cylinder and remains there until the next combustion. The actuation of the exhaust valve must be done depending on the desired EGR rate and the speed. The rigid actuation via a cam on the camshaft must therefore be replaced by a variable valve control. This is associated with considerable additional costs, which are relativized by the fact that the variable valve control enables further functions, for example, performance and consumption can also be optimized for the respective operating condition through variable control times [Basshuys17].

Suppliers like Bosch already offer a **water injection** that removes heat from the combustion chamber and thus counteracts the formation of NO_x [Sher98]. In contrast to exhaust gas recirculation, its influence on soot formation is small and not necessarily disadvantageous [Hountalas07]. The consumption is a few ml/km. However, it leads to additional system costs and becomes economical when it makes an EGR superfluous. A weaker form is the humidification of the intake air, but this is unlikely to be sufficient to replace an EGR. Since 2022, requirements for the injection water have been standardized [ISO31120]. Methods for extending the refill intervals of the water tank (e.g. condensate recovery from the air conditioning system) are still in the early stages of research. An alternative to water injection is the addition of water to the fuel with the same effect, but this addition does not allow switching off or variable dosing, in addition, components of the fuel injection must be designed so that they do not suffer damage from the poorer lubricating properties of a water-containing fuel. Also, with a high water content of the fuel after a longer standstill, a microbiological infestation is likely.

It is conceivable to use **materials** for components such as the EGR actuator, whose surface makes it difficult for soot to deposit and thus sooting. However, this requires a high research effort, also the component costs would be higher.

An early cleaning with commercially available sprays or otherwise may possibly prevent the sooting of the exhaust gas recirculation actuator, but for many vehicles the work effort for this is considerable. A well-accessible, service-friendly exhaust gas recirculation is difficult to reconcile with the goal of protecting the exhaust gas recirculation from temperature influences.

It is conceivable to **burn off** sooted EGR components similar to a particulate filter. This would no longer require as much attention to the temperature of the recirculated exhaust gas as before, but the effort for both development and installation is likely to be considerable. More realistic, but currently still in the experimental stage, is a **catalytic filter** in the exhaust gas recirculation. Simple mechanical filters are already common in low-pressure exhaust gas recirculation.

A **concentrated, engine-near exhaust aftertreatment**, perhaps even additional heating, can lead to more favorable operating temperatures and thus higher conversion rates, but in this case further measures against overheating must also be taken.

Many previously discussed measures improve the EGR. However, it should not be forgotten that ongoing improvements in combustion processes, which have been the subject of engine research for decades, can have a major impact that goes beyond the listed EGR-related measures. This can involve optimizations of known combustion processes, novel combustion processes like HCCI [Basshuys17] or completely unknown combustion processes. In the best case, better combustion processes make EGR unnecessary. Lower compression can also reduce NO_x emissions, but contradicts many other development goals for the diesel engine. A **variable compression**, which can resolve such goal conflicts, is however enormously complex if it is realized through a mechanical adjustment of the crank mechanism or even the cylinder housing. Variable control of the intake valve can also influence compression.

Exhaust aftertreatment by SCR can be effectively improved with a sufficiently large additive tank. It makes sense to combine the SCR with a storage catalyst that operates at lower temperatures or with a second, engine-near SCR system.

In addition to vehicle improvements the influence of the fuel should also be considered [Stein02, Song07]. Potentials are particularly expected from oxygen-containing fuels (e.g. alcohols, ethers). Alcohols could be, for example, ethanol or methanol; well-studied ethers with emission advantages are polyoxymethylene dimethyl ethers (OME) [Härtl14].

References

[Basshuys17]   van Basshuysen, R., Schäfer, F.: *Handbuch Verbrennungsmotor: Grundlagen, Komponenten, Systeme, Perspektiven.* 8th Ed., Springer-Vieweg, Wiesbaden (2017)

[BMVI16]   Bundesministerium für Verkehr und digitale Infrastruktur: *Bericht der Untersuchungskommission „Volkswagen" (2016).* https://www.kba.de/DE/Themen/Marktueberwachung/Abgasthematik/erster_ber_uk_vw_nox.pdf; jsessionid=C7C1342F0E39C5260D67B4881C657DFF.live21321?__blob=publicationFile&v=2. Accessed: 13. Febr. 2023

[Borgeest20E]   Borgeest, K.: *Elektronik in der Fahrzeugtechnik*, 4th Ed., Springer-Vieweg, Wiesbaden (2020)

[BR22]   Bayerischer Rundfunk, Meyer-Fünffinger, A., Streule, J.: *Neue Dokumente: Bosch und der Diesel-Skandal*, Bayerischer Rundfunk (17.11.2022). https://www.br.de/nachrichten/wirtschaft/neue-dokumente-bosch-und-der-diesel-skandal,TNLJeNz. Accessed: 13. Febr. 2023

[Cabraser16]   Cabraser, E.J.: Statement of claim. http://www.cand.uscourts.gov/filelibrary/1709/Consolidated_Consumer_Class_Action_Complaint.pdf.   Accessed: 24. Jan. 2023

[Domke15]   Domke, F., Lange, D.: *The exhaust emissions scandal („Dieselgate")*, 32C3 (2015), Video: https://www.youtube.com/watch?v=d9HJw3AUvGk, Slides: https://fahrplan.events.ccc.de/congress/2015/Fahrplan/system/event_att achments/attachments/000/002/812/original/32C3_-_Dieselgate_FINAL_s lides.pdf. Accessed: 13. Febr. 2023

[Domke16]   Domke, F.: Expert hearing about „Funktionsweise und Möglichkeiten von Abschalteinrichtungen und sonstigen Manipulationen einer NO_X-Abgasreinigung", positon to decision SV-3 for the session of the 5th committee of the 18th election period of the German Parliament, 22.09.2016.   https://www.bundestag.de/blob/461980/6222a71b56a8c71 0d7d9523c59e5e402/stellungnahme_domke-data.pdf. Accessed: 20. Febr. 2023

[Domke20]   Domke, F.: *Mercedes E350T Abgassystemanalyse* (2020). https://www.duh.de/fileadmin/user_upload/download/Pressemitteilungen/Verkehr/211104_DaimlerDieselgate/Gutachten_Abgasmanipulationen_de_geschwärzt.pdf. Accessed: 13. Febr. 2023

[DUH16]   Deutsche Umwelthilfe: *Deutsche Umwelthilfe verklagt Fiat-Chrysler Deutschland wegen irreführender Werbeaussagen zum Fiat 500x 2.0 beim Landgericht Frankfurt*, Press Release 06.07.2016. https://www.duh.de/presse/pressemitteilungen/pressemitteilung/deutsche-umwelthilfe-verklagt-fiat-chrysler-deutschland-wegen-irrefuehrender-werbeaussagen-zum-fiat-5. Accessed: 24. Jan. 2023

[D18/13118]   Deutscher Bundestag: *Unzulässige Abschalteinrichtungen in weiteren Audi-Fahrzeugen*, Circular 18/13118 of 14.07.2017. http://dipbt.bundestag.de/doc/btd/18/131/1813118.pdf. Accessed: 13. Febr. 2023

[ECE83]   Regulation No 83 of the Economic Commission for Europe of the United Nations (UN/ECE) – *Uniform provisions concerning the approval of vehicles with regard to the emission of pollutants according to engine fuel requirements*, Rev. 5 (2015). https://www.unece.org/fileadmin/DAM/trans/main/wp29/wp29regs/R083r5e.pdf + 14 Amendments (2016–2022). Accessed: 13. Febr. 2023

[EPA15]   EPA. https://www.epa.gov/sites/production/files/2015-11/documents/vw-nov-2015-11-02.pdf. Accessed: 20. Febr. 2023

[EPA98/08]   EPA. https://www.epa.gov/sites/production/files/2014-06/documents/defeat.pdf. Accessed: 20. Febr. 2023

[EuGH20]   EuGH, Case C-693/18, Judgement of 17.12.2020

[EU05/55]   *Directive 2005/55/EG of the European Parliamant and the Council of 28 September 2005 on the approximation of the laws of the Member States relating to the measures to be taken against the emission of gaseous and particulate pollutants from compression-ignition engines for use in vehicles, and the emission of gaseous pollutants from positive-ignition engines fuelled with natural gas or liquefied petroleum gas for use in vehicles.* https://eur-lex.europa.eu/legal-content/EN/TXT/PDF/?uri=CELEX:32005L0055. Accessed: 13. Febr. 2023

[EU08/692]   *Commission Regulation (EC) No 692/2008 of 18 July 2008 implementing and amending Regulation (EC) No 715/2007 of the European Parliament and of the Council on type-approval of motor vehicles with respect to emissions from light passenger and commercial vehicles (Euro 5 and Euro 6) and on access to vehicle repair and maintenance information.* https://eur-lex.europa.eu/legal-content/EN/TXT/PDF/?uri=CELEX:32008R0692. Accessed: 13. Febr. 2023

[EU11/582]   *Commission Regulation (EU) No 582/2011 of 25 May 2011 implementing and amending Regulation (EC) No 595/2009 of the European Parliament and of the Council with respect to emissions from heavy duty vehicles (Euro VI) and amending Annexes I and III to Directive 2007/46/EC of the European Parliament and of the Council.* https://eur-lex.europa.eu/legal-content/EN/TXT/PDF/?uri=CELEX:32011R0582, with subsequent changes: https://eur-lex.europa.eu/legal-content/EN/TXT/PDF/?uri=CELEX:02011R0582-20221210. Accessed: 13. Febr. 2023

[EU12/459]   *Commission Regulation (EU) No 459/2012 of 29 May 2012 amending Regulation (EC) No 715/2007 of the European Parliament and of the Council and Commission Regulation (EC) No 692/2008 as regards emissions from light passenger and commercial vehicles (Euro 6).* https://eur-lex.europa.eu/legal-content/EN/TXT/PDF/?uri=CELEX:32012R0459. Accessed: 13. Febr. 2023

[EU14/136]   *Commission Regulation (EU) No. 136/2014 of 11 February 2014 amending Directive 2007/46/EC of the European Parliament and of the Council, Commission Regulation (EC) No 692/2008 as regards emissions from light passenger and commercial vehicles (Euro 5 and Euro 6) and Commission Regulation (EU) No 582/2011 as regards emissions from heavy duty vehicles (Euro VI).* https://eur-lex.europa.eu/legal-content/EN/TXT/PDF/?uri=

CELEX:32014R0136. Accessed: 13. Febr. 2023

[EU18/1832] *Commission Regulation (EU) No 2018/1832 of 5 November 2018 amending Directive 2007/46/EC of the European Parliament and of the Council, Commission Regulation (EC) No 692/2008 and Commission Regulation (EU) 2017/1151 for the purpose of improving the emission type approval tests and procedures for light passenger and commercial vehicles, including those for in-service conformity and real-driving emissions and introducing devices for monitoring the consumption of fuel and electric energy.* https://eur-lex.europa.eu/legal-content/EN/TXT/PDF/?uri=CELEX:32018R1832. Accessed: 13. Febr. 2023

[EU98/69] *Directive 98/69/EC of the European Parliament and the Council of 13 Oktober 1998 relating to measures to be taken against air pollution by emissions from motor vehicles and amending Council Directive 70/220/EEC.* https://eur-lex.europa.eu/LexUriServ/LexUriServ.do?uri=CONSLEG:1998L0069:19981228:EN:PDF. Accessed: 13. March. 2023

[FAZ16] FAZ: *Amerikaner entdecken neue Betrugssoftware bei Audi* (06.11.2016). https://www.faz.net/aktuell/wirtschaft/unternehmen/umweltbehoerde-amerikaner-entdecken-neue-betrugssoftware-bei-audi-14514808.html. Accessed: 13. Febr. 2023

[Führ16] Führ, M.: *Gutachterliche Stellungnahme für den Deutschen Bundestag – 5. Untersuchungsausschuss der 18. Wahlperiode,* fortgeschriebene Fassung (19.11.2016). https://www.bundestag.de/resource/blob/481344/c6f582c85 98c9d6b62fcfb2acd012462/stellungnahme-prof--dr--fuehr--sv-4--data.pdf. Accessed: 23. Febr. 2023

[Härtl14] Härtl, M., Seidenspinner, P., Wachtmeister, G., Jacob, E.: *Synthetischer Dieselkraftstoff OME1 — Lösungsansatz für den Zielkonflikt NOx-/Partikel-Emission.* MTZ – Motortechnische Zeitschrift. 75. pp. 68–73 (2014)

[Hauff13] Hauff, K.: *Thermische Alterung und reversible Deaktivierung von Dieseloxidations- und NO_x-Speicherkatalysatoren.* Dissertation, Universität Stuttgart, Logos-Verlag, Berlin (2013)

[Hountalas07] Hountalas, D., Mavropoulos, G., and Zannis, T.: *Comparative Evaluation of EGR, Intake Water Injection and Fuel/Water Emulsion as NO_x Reduction Techniques for Heavy Duty Diesel Engines,* SAE Technical Paper 2007-01-0120 (2007)

[Hüning19] Hüning, F.: *Nachrüstungsmöglichkeiten von Dieselfahrzeugen aus technischer Sicht,* NZV 27(2019)

[ISO15031-3] TC22, SC31: *Road vehicles – Communication between vehicle and external equipment for emissions-related diagnostics – Part 3: Diagnostic connector and related electrical circuits, specification and use,* 3rd Ed. ISO 15031-3:2023

[ISO22241] TC22, SC34: *Diesel engines – NO_x reduction agent AUS 32,* ISO 22241, Parts 1 … 5 (2017 … 2019)

[ISO27145] TC22, SC31: *Road vehicles – Implementation of World-Wide Harmonized On-Board Diagnostics (WWH-OBD) communication requirements,* ISO 27145, Parts 1–6 (2012 … 2023)

[ISO31120]   TC22, SC34: *Road vehicles — Injection water — Part 1: Quality requirements*, ISO 31120-1:2022

[Kolar90]   Kolar, J.: *Stickstoffoxide und Luftreinhaltung*. Springer, Berlin, Heidelberg (1990)

[Kroher16]   Kroher, T., Rattay, U.: *Sauber gemacht*, ADAC Motorwelt 7/8, p. 38 (2016)

[Monitor16]   WDR, Monitor, Press Release (17.05.2016). https://www1.wdr.de/daserste/monitor/extras/opel-134.html. Accessed: 28. Jan. 2023

[MSE]   Mercedes-Schadenersatz. https://mercedes-schadensersatz.de/magazin/slipguard-und-bit15-abschalteinrichtungen-entdeckt. Accessed: 13. Febr. 2023

[Opel16]   Adam Opel AG: Position (20.05.2016). http://media.opel.de/media/de/de/opel/home/news-archive.month.05.year.2016.html. Accessed 20.12.2016, not longer available.

[Pott97]   Pott, E.: *Verfahren und Vorrichtung zur Überwachung der De-Sulfatierung bei NO_x-Speicherkatalysatoren*, patent application DE19731624A1, filed 23.07.1997

[Pott98]   Pott, E., Gloger, J., Bosse, R.: *Schwefelfalle und De-Sulfatierung des Abgasreinigungssystems einer Brennkraftmaschine*, patent application DE19855089A1, filed 28.11.1998

[Schwering20]   Schwering, A.: *Abgasskandal im Getriebe-Modus: ZF zahlt 4,2 Mio. Bußgeld – AL951 & AL551 zu leicht manipulierbar*. Anwalt.de (2020). https://www.anwalt.de/rechtstipps/abgasskandal-im-getriebe-modus-zf-zahlt-42-millionen-euro-bussgeld-al951-zu-leicht-manipulierbar_179398.html. Accessed: 18. Febr. 2022

[Sher98]   Sher, E.: *Handbook of Air Pollution from Internal Combustion Engines: Pollutant Formation and Control*. Academic Press, San Diego CA (1998)

[Song07]   Song, C.-L., Zhou, Y.-C., Huang, R.-J., Wang, Y.-Q., Huang, O.-F., Lü, G., Liu, K.-M.: *Influence of ethanol-diesel blended fuels on diesel exhaust emissions and mutagenic and genotoxic activities of particulate extracts*. J. Hazard. Mater. **149**(2), 355–363 (2007)

[Stein02]   Stein, H., Krahl, J., Munack, A., Schröder, O., Dutz, M.: *Measurement of particle emissions from diesel engines operating on different fuels*. Landbauforschung Völkenrode. Special Ed. **235**, 95–101 (2002)

[Stein16]   Stein, S.: *Abgasrückführung und Ladedruckregelung eines Dieselsteuergerätes (EDC)*. Master Thesis, Aschaffenburg University (2016)

[Traufetter21] Traufetter, G.: *Daimler blitzt in Flensburg mit Widerspruch gegen Megarückruf ab*. Spiegel-Online, 05.02.2021. https://www.spiegel.de/auto/daimler-blitzt-im-abgasskandal-in-flensburg-mit-widerspruch-gegen-megarueckruf-ab-a-3d791add-d8fe-403f-9418-86f3b78267fd. Accessed: 28. Jan. 2023

[Verbraucher]   GoldenLegalTech Services GmbH: https://verbraucherhammer.de/themen/abgasskandal-entschaedigung. Accessed: 28. Jan. 2023

[VG_SL]   VG Schleswig, 3 A 113/18, Judgement of 20.02.2023

[Welt17]   Fabricius, M., Vetter, Ph.: *Nach Dieselgate droht jetzt der Dezibel-Skandal*, Welt, 09.08.2017. https://www.welt.de/wirtschaft/article167543376/Nach-Dieselgate-droht-jetzt-der-Dezibel-Skandal.html. Accessed: 13. Febr. 2023

[Wienand06]   Wienand, Kh., Ullrich, K., Muziol, M.: *Abgasrückführung mit einem Anemometer*, patent application DE102006058425A1, filed 08.12.2006

[WiWo16/10] Reimann, A.: *VW-Kunden klagen über Probleme nach Abgas-Rückruf*,
            Wirtschaftswoche, p. 40 (4 Oct 2016)

Law 5

5.1 Terms

Environmental law distinguishes between emission and immission. Emission is the release of pollutants, noise or radiation, here specifically the production of harmful exhaust gases by vehicles. Immission is the resulting impact of pollutants, noise or radiation on humans, animals, plants, and objects (especially buildings). Effects on the atmosphere (e.g., greenhouse effect) are usually not covered by the term immission, but are no less significant.

Many Citizens underestimate the influence of the EU on legislation. Especially in exhaust gas legislation, member states have little room for manoeuvre, this is mainly done in Brussels and not in Berlin. The secondary law of the EU (in contrast to primary law, which defines the internal structure of the EU) knows two essential types of legal sources, the *Directives* and the *Regulations* (engl. Regulations). Although some directives and regulations cover over 100 pages, the reader is advised to look at the cited documents, all of which are available as PDF documents on the EU server (see literature references in the book), as their uniform structure allows for quick orientation. Directives and regulations consist of three parts, a brief justification, the actual text and the appendices, with the appendices often contributing most to the volume, as technical details, e.g., measurement methods and terms are described there. The directives are to be implemented by the member states into national law, European regulations have direct effect throughout the EU. The implementation of European directives in Germany is done through *laws* and implementing *regulations* (the term "regulation" has a different meaning in German law than in European law).

If a member state does not comply with EU law, the EU Commission or another member can initiate *infringement proceedings* according to Art. 258 ff.

© The Author(s), under exclusive license to Springer Fachmedien Wiesbaden GmbH, part of Springer Nature 2025
K. Borgeest, *Manipulation of Exhaust Gas Values*,
https://doi.org/10.1007/978-3-658-45864-5_5

[TFEU]. First, the EU Commission writes to the member (the "blue letter from Brussels"), if there is no appropriate response, the case is brought before the European Court of Justice (ECJ).

German legal sources are laws, regulations, statutes and administrative regulations. Laws are created in a democratic process, therefore require a parliamentary majority; subordinate legal sources, on the other hand, can be decided by ministries or authorities. Since it is not sensible to burden a parliament with technical details, laws contain few, generally formulated principles, which are detailed by regulations or other sources of law. In this case, a law refers to details elsewhere. This book occasionally refers to federal laws and regulations, in other areas of law that are not relevant here, the competence to legislate or to decide on other sources of law lies with the federal states or local authorities. Although almost all areas of public and partly also private life are regulated by the legislator in Germany and the volume of legal sources is increasing, there are always smaller or larger gaps, in particular not every individual case can be regulated by law in advance. These are filled by *jurisprudence*, but not independently, but on the occasion of a lawsuit. A judgment thus issued almost always, but not necessarily, serves as a template for other judgments, thus judgments in precedent cases practically have a similar effect as legal norms, although they are not. The judiciary is divided into various areas, not all of which are relevant for the topic of the book. Relevant areas are civil jurisdiction (if, for example, a buyer of a manipulated car buyer sues the seller or manufacturer for damages) and criminal jurisdiction (which, for example, after successful investigations by the public prosecutor's offices, finds the offenders responsible for the exhaust gas scandal guilty and determines their sentence in the judgment). The most important source of civil law is the Civil Code [BGB], the most important source of criminal law is the Penal Code [StGB], especially in connection with the exhaust gas scandal, however, criminal provisions in other sources of law (secondary criminal law) are also relevant. The corresponding pre-trial and court proceedings are regulated in the Code of Criminal Procedure [StPO] and the Code of Civil Procedure [ZPO]. Administrative jurisdiction is also relevant, for example, when an affected citizen or an environmental association sues for compliance with immission values.

5.2   Type Approval

In addition to the individual vehicle, each new vehicle type must be approved; this type approval, which is significantly more complex than the registration of a single new car, is also called homologation. Its purpose is to ensure compliance with legal regulations on the vehicle side, especially in the two areas of emissions and safety. In terms of emissions, the verification of the already presented exhaust emission limits according to the also presented procedures represents the largest temporal and financial aspect. Noise emissions are also important. [EU91/441] introduced five different tests for car emissions, which are defined in their current version in [ECE83]:

- Type I: "Verifying the average exhaust emissions after a cold start",
- Type II: "Carbon monoxide emission at idling speed", initially only for a few cars, extended to all cars without a diesel engine in [EU08/692],
- Type III: "Emission of crankcase gases" in petrol engines,
- Type IV: "Evaporation emissions" in petrol engines,
- Type V: "Durability of anti-pollution devices" with exceptions specified in the directive,
- Type VI (introduced subsequently): "Verifying the average low ambient temperature carbon monoxide and hydrocarbon exhaust emissions after a cold start", not applicable to diesel engines, but manufacturers must demonstrate according to [EU08/692] that the NO_x aftertreatment device reaches a sufficiently high temperature for proper operation within 400 s after a cold start at $-7\ ^\circ C$, as in the Type VI test.

In addition, the electronic diagnosis of exhaust-relevant errors in the vehicle (OBD) must be checked, which can be counted as a seventh check.

For the emission scandal, Type I as the core of the tests according to the cycles discussed in Sect. 3.1 and Type VI are relevant. [EU08/692] also requires the measurement of CO_2 consumption, a check of the conformity of vehicles in operation (in-service conformity) and a check of the on-board diagnosis (Sect. 4.6). Table I.2.4 in [EU14/136] shows in detail which tests apply to which engines. After type approval, it must be demonstrated that the produced vehicles comply with the type approval (CoP, Conformity of Production) [EU17/1151, ECE83]. In addition to the CoP, proof of "conformity in operation" is also required.

A type approval carried out in one EU country is usually recognized by other EU countries, so a manufacturer will not carry out this procedure in every EU country, but only in the country that appears most favorable from the manufacturer's point of view. In addition to the lowest possible test severity, other criteria such as costs may also play a role. Type approvals in Luxembourg and Malta are noticeably frequent [Rogert16].

Alternatively to the EU approval of vehicles or vehicle components, ECE approval (Economic Commission for Europe of the UN) is possible. EU directives and regulations then refer in the annex to accepted ECE rules; increasingly, the ECE rules become part of EU legislation by EU legal norms referring without alternative to ECE rules or parts thereof. With regard to exhaust emissions, these are the ECE rules [ECE24] (visible pollutants from diesel engines), [ECE40] (visible pollutants from motorcycles), [ECE47] (exhaust gases from mopeds), [ECE49] (exhaust gases from gas engines) and [ECE83] (emissions). With regard to CO_2 emissions, [ECE84] (fuel consumption) and [ECE101] (measurement of CO_2 and fuel consumption) are relevant, in a broader sense also [ECE85] (engine power). An EU framework directive on the type approval procedure, which also refers to technical individual directives, is [EU07/46].

Members of the ECE agreement include not only EU countries but also other European countries and now also some non-European countries, but the practice of mutual recognition is not as strictly adhered to in all countries as in the EU. Japan is also a member of the agreement, but only recognizes ECE approvals in a few areas.

Regardless of the country and whether in the EU or ECE system, type approval consists of two parts:

- an examination of the vehicle type, which is not carried out by state institutions, but by state-recognized private organizations (e.g., TÜV), and
- the official approval procedure based on the test report.

The private testing facility is commissioned and paid for by the vehicle manufacturer for the testing service. In individual cases, the private type testers provide additional commercial services for their customers. Possible alternatives to avoid a mix of commercial and public interests would be the official commissioning of the testing institute or the testing by the authorities themselves.

A vehicle with an unauthorized defeat device does not meet the approval requirements, so the operating permit usually expires if this is subsequently discovered [Führ16]. Among lawyers, there are different views on whether the KBA must revoke unlawful approvals or whether the operating permit of the manipulated vehicles automatically expires. In recent years, the KBA has called on owners, for whose vehicles it ordered retrofits (almost always in the form of software updates), to carry out these measures; only if these retrofits were not carried out within the set deadline, the operating permit is withdrawn.

5.3 Immissions

The impact of exhaust gases, e.g., from the operation of vehicles in non-public, operational facilities, is not subject to environmental law, but to occupational safety. Technical Rules for Hazardous Substances 554 [TRGS554] are particularly relevant for diesel exhaust fumes in the workplace.

The immissions imposed to the citizen are defined in the EU by directives that are implemented by the member states, in Germany this is the Federal Immission Control Act, which is detailed by almost 40 regulations and supplemented by the Technical Instructions for Air Pollution Control [TALuft]. Since emissions and immissions are causally linked, they are not strictly separated in German immission legislation. In connection with exhaust gases from combustion engines, the following regulations on the Federal Immission Control Act (short BImSchV) are particularly relevant:

- 10. BImSchV, Regulation on the quality and labeling of fuel and combustion materials [BImSchV10],
- 13. BImSchV, Regulation on large combustion, gas turbine and internal combustion engine plants [BImSchV13],
- 28. BImSchV, Regulation on emission limits for internal combustion engines [BImSchV28],
- 35. BImSchV, Regulation on the labeling of vehicles with low contribution to pollutant load [BImSchV35],
- 39. BImSchV, Regulation on air quality standards and maximum emission levels [BImSchV39],
- 44. BImSchV, Regulation on medium-sized combustion, gas turbine and internal combustion engine plants [BImSchV44].

The 10th BImSchV defines the permissible sulfur content, which is responsible for the emission of sulfur oxides, according to the European Directive [EU09/30] and its predecessor directives; it also sets further requirements for fuel quality and contains procedural rules. The 13th BImSchV applies to engines that do not serve as drives, e.g., for power generation, and implements, among other things, the Directive [EU01/80]. The 28th BImSchV essentially consists of a reference to the EU Directive [EU16/1628] and associated directives; it concerns engines in mobile machines and devices. It also includes small engines for garden tools, which the EU allows particularly high emissions. Due to the long period of use in some applications, legal adjustments have a longer delay than in road traffic.

The thirty-fifth BImSchV introduces traffic bans (environmental zones) in Germany which are unique in Europe, as well as badges for exemption from traffic bans under certain conditions. The 39th BImSchV defines the permissible pollution of the air with pollutants, i.e., immissions in the narrower sense, as well as procedures for their assessment. It also takes into account particles and nitrogen oxides, with NO_2 being considered separately from other nitrogen oxides. It mainly implements the directives [EU01/81, EU04/107, EU08/50]. The 44th BImschV implements EU law for non-approval-required stationary plants.

For each pollutant emitted in significant quantities by vehicles, the legislator regulates which quantities may be released under which conditions, as well as measurement procedures (Chap. 3) for determining these quantities.

5.4 Emissions

Emission law is most significant for the exhaust scandal. Since violations in the USA have more serious consequences due to functioning surveillance as well as higher penalties and compensations, the emission law outside Europe, especially in the USA, will also be briefly discussed in the following.

5.4.1 European Union

As already described in the case of immission law, emission law also consists of directives and regulations of the EU. The scope for own legislation by the nation states is small, they can and must decide how compliance with regulations is monitored and how violations are punished. The EU has tightened the exhaust emission limits since their introduction in several stages.

For the type approval of two-wheelers, trikes, and quads, the Euro 5 emission standard has been in place since 2020 according to the original plan, and since 2021 for all individual first registrations (however, the industry was able to use the Covid-19 pandemic as a reason for a postponement [EU20/1694]). In addition to the limit values, a significant innovation is that the previous distinction between light motorcycles and motorcycles in emission law has been abolished. The NO_x limit is 90 mg/km. The particle mass is limited to 4.5 mg/km, the particle number is still unlimited in contrast to cars. Other emissions besides NO_x and particles are also regulated. Directive 97/24/EC [EU97/24] originally set all requirements for type approval, including emissions. The directive has since been updated with regard to emissions by many subsequent directives [EU02/51, EU03/77,

EU05/30, EU06/72, EU06/120, EU09/108, EU13/60] and regulations [EU13/168, EU14/134, EU16/1824, EU18/295, EU21/1445]. Regulation 134/2014 with two subsequent regulations is interesting because, like cars, it introduces a new test cycle. The test definition is shifted from EU legal norms, as with cars, to ECE documents. For two-wheelers, this is [ECE_GTR2], which means ECE 40 and ECE 47 lose their original significance.

For passenger cars with petrol and diesel engines, the Euro 6 standard has been in place since 2014 (type approval)/2015 (first registration) (Euro 7 is expected in 2025). The emission scandal also affects older vehicles. Table 5.1 and 5.2 show the NO_x and particle limit values for cars. The decisive factor for the values is **Regulation 715/2007** [EU07/715]. Compared to other European legal sources, it is very short with 16 pages including the annexes; it is one of the most important documents in relation to the emission scandal. It has been updated in the meantime by the regulations [EU08/692, EU09/595, EU11/566, EU12/459, EU17/1151, EU17/1154, EU18/858 EU18/1832], in particular, Euro 6 was divided into several sub-stages 6b to 6d (6a never existed). The same limit values apply in principle to these sub-stages, differences lie in the test procedure:

- Euro 6b: NEFZ (section 3.1.1),
- Euro 6c: WLTP (Sect. 3.1.2),
- Euro 6d-TEMP: WLTP + RDE with generously allowed exceedances of the limit values at RDE (Sect. 3.1.3) and
- Euro 6d like 6c with less generously allowed exceedances,
- Euro 6e like 6d with further reduced exceedances.

Table 5.1 NO_x- and particle limit values for passenger cars with petrol engines in the EU

Standard	Euro 4	Euro 5	Euro 6
Introduction type approval	01.01.2005	01.09.2009	01.09.2014
NO_x in mg/km	80	60	60
Particle mass in mg/km (direct injection only)		4,5	4,5
Particle number per km (direct injection only)			$6 \cdot 10^{12}$, from 01.09.2017/2018 (type approval/first registration) $6 \cdot 10^{11}$

Table 5.2 NO_x- and particle limit values for passenger cars with diesel engines in the EU

Standard	Euro 4	Euro 5a	Euro 5b	Euro 6
Introduction type approval	01.01.2005	01.09.2009	01.09.2011	01.09.2014
NO_x in mg/km	250	180	180	80
Particle mass in mg/km	25	5	4,5	4,5
Particle number per km			$6 \cdot 10^{11}$	$6 \cdot 10^{11}$

From September 2023 type approvals are possible for Euro 6e only (first registrations 1 year later) [EU23/443].

Since direct injectors emit significantly more particles than older petrol engines, a particle limit was introduced for the first time with Euro 5. Since the number of particles is more important for their health hazard than their mass, the number of particles is limited from Euro 6 onwards. It is noteworthy that initially a particle number 10 times higher than that of diesel engines was permissible. The automotive industry enforced this regulation in order to save a particle filter.

A limit for the number of particles was introduced for diesel engines for the first time with Euro 5b.

For commercial vehicles, the Euro VI emission standard applies according to [EU07/715, EU11/582, EU09/595] (in contrast to the car standards designated with Roman numerals).

The mentioned legal sources also specify under which conditions the limits apply. Here again [EU07/715] is interesting. The most discussed paragraph of European legal sources is Art. 5 (2). It was created with the directive [EU98/69] in reference to US legislation, but there are strict requirements for the justification of shutdown devices in contrast to the EU.

Literal quotation from Art. 5 (2) of Regulation 715/2007

"The use of defeat devices that reduce the effectiveness of emission control systems shall be prohibited. This prohibition shall not apply where:

(a) *the need for the device is justified in terms of protecting the engine against damage or accident and for safe operation of the vehicle;*
(b) *the device does not function beyond the requirements of engine starting;*

or

(c) *the conditions are substantially included in the test procedures for verifying evaporative emissions and average tailpipe emissions."*

This article establishes the illegality of many defeat devices. Since the engine is in a difficult operating condition during start-up (b.), emission control systems only need to work fully after starting. A device that would be required in the test procedures for Type I or Type VI (c.) cannot be illegal either.

Many manufacturers try to justify violations by referring to the engine protection mentioned under (a.). The Scientific Services of the Bundestag and other lawyers consider their principle of making the exception the rule, for example in the form of narrow temperature windows, based on the unclear formulation, to be illegal [Bundestag16, Führ16]. The Federal Ministry of Transport relativized such a clear assessment after the censorship of its report by some manufacturers [dpa16] [BMVI16]. The ECJ [ECJ20] clarified the ban: It is legal to prevent an imminent, sudden damage, but not to use defeat devices instead of a durable design. With this judgment, the argumentation of many manufacturers collapsed. Another judgment [ECJ22] largely refers to this and formulates:

> *"The clogging up or the ageing of the engine cannot, in any event, be regarded as an 'accident' or 'damage', within the meaning of that provision, since such occurrences are, in principle, foreseeable and inherent in the normal operation of the vehicle."*
> and *"The prohibition laid down in Article 5(2) of that regulation would be devoid of substance and deprived of any effectiveness if car manufacturers were permitted to equip motor vehicles with such defeat devices with the sole aim of protecting the engine against clogging up and ageing."*

A clear formulation of Article 5 would not have brought any economic disadvantages, as it affects all market participants in the same way; the position of some lobbyists that this would result in competitive disadvantages is not comprehensible. On the contrary, compliance with the regulation would have resulted in a technical advantage.

So far, the verification of compliance with the limit values at type approval has been discussed. Another question is how compliance with exhaust emission limits is ensured during the vehicle's operation. Putting every vehicle on a vehicle test stand for type approval at regular intervals would answer this question, but is not feasible for cost reasons. This would be even more expensive for trucks; since type approval there is not carried out on the vehicle test stand, but on the engine test stand, the engine would have to be regularly removed and placed on an engine test stand. The legislator has therefore defined more generous monitoring limits in addition to the approval limits, which are much cheaper to check. With each new emission standard, new monitoring limits are usually also defined, with the monitoring limits gradually approaching the approval limits in the long-term trend (Sect. 4.6).

Euro 7 will probably come into force in 2025. In addition to tightening, for example, the limit values for nitrogen oxides and particles, exhaust components previously irrelevant for legislation in the EU are also likely to be limited. This can affect, for example, ammonia (NH_3), nitrous oxide (N_2O) and formaldehyde (CH_2O), which are already partly regulated in some regions outside Europe. Methane (CH_4) is no longer included in the latest proposal of the EU Commission. Concerning particles, there will be a separate limit value for ultrafine particles. As these are new substances to be measured, there will be significant changes in the measurement technology behind the scenes, and experiences from other regions where regulation already exists can be drawn upon. Particles from brakes and tyre wear are also to be regulated. Since these are composed differently than carcinogenic diesel exhaust gases, separate limit values make sense. This also affects electric vehicles, although these usually brake electrically and therefore do not release brake wear. The durability of emission-reducing devices is to be increased. By eliminating restrictive conditions, the RDE will be adapted even more closely to reality. Organisationally, there will probably be a common set of rules for cars, motorcycles and commercial vehicles in the future.

5.4.2 Countries outside the EU

Switzerland has been applying limit values according to Sect. 5.4.1 since 1995, in accordance with the relevant regulation there [TAFV]. In the United Kingdom, EU rules applied until 31.12.2020; it is becoming apparent that even after the British exit from the EU, its emission legislation will continue to be applied there.

1963 under the responsibility of the US Environmental Agency EPA (Environmental Protection Agency) in the USA the Clean Air Act (CAA) [EPA] came into force, a very comprehensive air pollution control law. In addition to minor revisions, major revisions regarding legal supervision took place in 1970 (first limits for immissions and vehicle emissions) and 1990 (stricter standards, tightened enforcement). The emission limits are regularly updated. For vehicles, Title II (Emission Standards for Moving Sources) is relevant.

In addition to the legislation under the supervision of the federal agency EPA, there is additional legislation from federal states, in this context the CARB (Californian Air Resource Board) is to be mentioned, which enforced stricter emission legislation in California. The Californian legislation was also adopted by other federal states. The Canadian legislation resembles the federal legislation in the USA.

The EPA standards provide for several temporal stages *(Tiers)*. Tier 3 is being introduced from 2017 to 2025. Each Tier is divided into several pollutant levels *(Bins)*. Emissions according to Tier 3 must be maintained over a lifetime of 150,000 miles. In addition, manufacturers must still comply with average limits over all vehicles sold in a year, which are continuously reduced according to a predetermined plan. This makes the EPA regulations considerably more complex than in the EU, these can be read for example in [BorgWarner22].

California and the states adopting Californian legislation have introduced the LEVIII (Low Emission Vehicle) stage. There too, there are several categories similar to the Bins of the EPA as well as continuously lowered fleet average values [BorgWarner22]. The EPA intends to adapt its standards to the previously stricter Californian standards in the future.

In the USA, stricter limits for pollutants apply than in Europe, however, consumption plays a lesser role there and thus also the CO_2 emissions. The goals there are mainly achieved with gasoline engines, the success of the Toyota Prius as the first freely available hybrid car on the US market may be due to the limits there.

Australia, China, India, and Turkey roughly follow EU legislation, South Korea follows US legislation for gasoline engines, otherwise mainly EU legislation. Other countries also predominantly use European or American standards, often in older versions [Dieselnet].

5.5   Consumer Rights

Consumer rights not only differ between Europe and the USA, but compared to emission legislation, they are also little harmonized within the EU. The greatest costs can arise for a manufacturer from claims in the USA.

In case of product defects in Germany, the legal liability for defects according to BGB of the commercial provider applies, which, unlike a guarantee, cannot be excluded by terms of business. For consequential damages from a product defect, product liability applies. The contract partner of the buyer is not the manufacturer of the vehicle, but the dealer.

To perceive the liability for material defects a material defect must be present according to § 434 BGB, this is not as obvious in the case of manipulation of emission values as it is with many other material defects. In the case of a material defect, the buyer can demand supplementary performance (e.g., in the form of a defect-free software update of the engine control unit), regardless of further claims for damages, possibly also cancel the purchase contract or reduce the purchase price. For the reduction (according to § 434 or also according to § 826 BGB), the term "small damages" has been established in the emission scandal, for the return the term "large damages". However, after an update, the original software is no longer available as evidence in a possible legal dispute. After half a year, the burden of proof for material defects is reversed, i.e., the buyer must be able to prove the defect. The liability for material defects is limited to two years, so it would have basically expired for the manipulations that became known in 2015 (cf. [OLG_M19]); for manipulations that were uncovered recently, a buyer should respond quickly. However, the LG Frankfurt/Main argues [LG_F19] that due to earlier case law it was not even possible in 2015 to actually claim the liability for material defects, and therefore the statute of limitations has not automatically expired. In addition to the liability for material defects, a claim according to § 852 BGB is also possible in the case of unlawful actions, which only becomes statute-barred after ten years. This paragraph shows that even the determination of whether a claim still exists is not trivial for the consumer and may require legal assistance. For contracts between merchants, the Commercial Code applies [HGB].

A fraudulent deception according to § 123 BGB, which is assumed in the case of a deliberately implemented function for manipulating emission values [LG_M16], allows the purchase contract to be contested. Regardless of claims for damages (esp. § 826 BGB, immoral intentional damage, possibly also § 823 BGB, liability for damages), the purchase contract can be contested up to one year after becoming aware in this case. The complication is that the deception is committed by the vehicle manufacturer or the engine manufacturer, but the contract partner of the buyer is the dealer, who usually did not know about the manipulations and, according to the OLG Celle, does not have to attribute them to himself [OLG_CE16]. However, a dealer who belongs to a manipulating automotive group is responsible [LG_M16]. In current proceedings due to fraudulent deception, the responsible manufacturer is the defendant, not the dealer.

After almost a year of unsuccessful lawsuits, there were many judgments, e.g., [LG_KR16], which strengthen the position of the consumer. The latter judgment allows the return of an affected vehicle in exchange for a refund of the purchase price minus compensation for the already occurred use, because even after the update the defect-free condition is not certain and the buyer can no longer trust the manufacturer. The LG Braunschweig found that the manipulation is not an insignificant defect, a period of more than three months for supplementary performance is unreasonably long, and also affirmed the return of the vehicle [LG_BS16]. The return of manipulated vehicles has now become established, usually a compensation for use is deducted from the refunded purchase price. This case law has also been adopted by the BGH [BGH252/19]. Although the compensation for use is widespread, there are also critical voices against it, as it creates an incentive for the defendant to unnecessarily delay a procedure. The return against compensation for use only has the character of a used car sale for the owner of a vehicle that has been used intensively and for a long time, and hardly any longer that of a real compensation. The LG Erfurt wanted to have the legality of a compensation for use clarified by the ECJ; VW tried to preempt this clarification with a payout to the plaintiff.

The BGH denied an additional claim of the injured party for delictual interest [BGH397/19].

A comprehensive overview of judgments on the emission scandal for scientific purposes is provided by [Heese], an overview of judgments in favor of consumers by [Warentest]. Initial lawsuits failed because plaintiffs could not prove the defects of their vehicle. Since for many affected vehicles there is still no recall by the Federal Motor Transport Authority KBA and also the parallel criminal proceedings, in which searches provide evidence, are only progressing slowly, more and more courts are commissioning expert opinions and increasingly demanding the defendant's cooperation in clarification (secondary burden of presentation). Seizure of documents as in criminal proceedings is not possible in civil proceedings in Germany.

The expert's report can, for example, include an experimental investigation on a roller test bench or an analysis of the software and the data set in the control unit. This results in costs that have now risen to almost €30,000 due to limited capacities (which are initially to be borne by the plaintiff). It is also a strategy of the defendants to drive up the costs of proving the case. It is much cheaper and faster if the expert can access information already known to him. The expert is usually appointed by the court. The parties can also commission experts; however, as these are not independent, some judges may view such private reports with skepticism. In fact, the author knows of a person (there may be others) who is paid by the lawyers of a vehicle manufacturer for false reports. Such a report then usually requires a counter-report from the other party for clarification and causes (perhaps intended by the initiators) a lot of confusion in the proceedings.

The lawyers of the defendant car companies try to push judges, who are expected to rule in favor of the plaintiff based on previous judgments, out of the proceedings due to bias. Although the German judiciary is largely immune to influence by companies, a judge at the Stuttgart Regional Court was replaced in all diesel proceedings by judge Sonn, who was previously a lawyer at the law firm Gleiss Lutz, which represents both Daimler and VW [Bender20].

In the USA, class actions are taking place against vehicle manufacturers, especially Volkswagen [Cabraser16]. It is in line with American customs to also systematically check the entire supply chain for possible lawsuits. A direct supplier was also affected by a class action [boschvw]. An American law firm known to the author initially also considered lawsuits against indirect suppliers, even against Infineon as the manufacturer of a microcontroller used in many control units and against ETAS as the manufacturer of tools for control unit development. Since microcontrollers and development tools are necessary to develop and build control units, but there is no specific connection with defeat devices, no such lawsuits were filed.

5.5.1 Collective Consumer Protection

Claims are in Germany to be individually registered and possibly legally enforced by each affected buyer; the risk of litigation costs initially lies with the buyer. German law does not provide for class actions by several buyers following the US model. This makes it considerably more difficult for the German consumer to assert his rights and floods judges and courts with similar proceedings. The legal service provider Myright, which litigated similar to a class action through assignment agreements, was stopped by the courts due to the specific design of the agreements ([LG_IN18] and others), a judgment of the Federal Court of Justice finally paved the way for collected lawsuits [BGH418/21], even though it concerned plaintiffs from Switzerland. At least VW was able to gain time with the action against legal service providers.

In 2018, the model declaratory action was introduced into German law (§§ 606–614 ZPO). It allows certain associations to legally clarify a disputed fact on behalf of any number of consumers. After a successful model declaratory action, however, each affected person still has to individually enforce his resulting claims. In the first model declaratory action in the emissions scandal, the Federal Association of Consumer Centres sued VW before the OLG Braunschweig because of the EA189 engines, this led to a settlement a few weeks later after VW initially let the settlement negotiations fail in February 2020. At the same court, another model declaratory action by the South Tyrol Consumer Centre is pending because of the EA189. At the OLG Stuttgart, a model declaratory action against Mercedes-Benz is pending over the models GLC 220d 4Matic, GLC 250d

4Matic, GLK 220 BlueTec 4Matic and GLK 250 BlueTec 4Matic. Consumers can inform themselves about the progress at the Federal Justice Office [BJA] and possibly also register in the lawsuit register, in addition, the suing associations usually publish ongoing information.

In 2020, despite resistance from German industry associations, there was a basic agreement in the EU on a directive that should introduce EU-wide association actions, similar to the class actions known so far from various countries, in all member states. Unlike the model declaratory action, it allows a possibility to directly enforce rights beyond the determination of a disputed fact. This directive was adopted and published in 2020 [EU20/1828], it should have been implemented into German law by the end of 2022 at the latest. Germany let the deadline pass due to different views of two ministries. As a result, among other things, an EU infringement procedure was initiated against Germany [EU23], but the disputing ministries have since agreed on a compromise.

5.5.2 Secondary Aspects

Due to the high value in dispute, high court costs can be expected. For the assessment, expenses in the five-figure range can arise. Initially it has not been clear that these expenses are covered by legal protection insurances. Thus, before the current wave of lawsuits against manipulating vehicle manufacturers, there was a smaller wave of lawsuits against some legal protection insurers who initially refused to cover the costs. By now, such refusals should be the exception. Some specialized law firms also offer solutions for victims who do not have legal protection insurance.

Another secondary aspect concerns indirect costs associated with a purchase contract that is invalid due to intentional damage. This mainly affects interest for credit financing. The best chances of reimbursement exist here with a car loan through a bank affiliated with the manufacturer, so legal advice can be useful here. An independent loan is unlikely to be unwound, but a refund of the costs by the vehicle manufacturer is conceivable.

5.6   Criminal Prosecution

The goal of criminal law is the punishment of the perpetrators, the societal benefit lies among other things in the deterrence of potential future perpetrators. Criminal law in Europe differs significantly from American criminal law, and even within the USA and the EU, criminal law is not uniform.

A significant difference to the USA is the American corporate criminal law. In addition to natural persons, legal entities can also be held accountable for crimes there, in Germany this is so far only possible for administrative offenses. The fines of recent years for administrative offenses including possible disgorgement in connection with the emissions scandal were also not negligible for the companies involved, e.g. VW 1 billion €, Daimler 870 million €, Audi 800 million €, Porsche 535 million €, Bosch 90 million €, Opel 64.8 million €, BMW 8.5 million € [Bender20, Flaig21, Lübben21]. Due to cartel-related illegal agreements between Daimler, BMW and VW with the aim of using undersized tanks for the urea solution together, fines to the EU (VW 502.4 million €, BMW 372.8 million €, Daimler as crown witness 0) are added [Mussler21].

The necessity of corporate criminal law in Germany is controversially discussed, especially in connection with the emissions scandal. A company like Volkswagen is not a criminal organization, but consists of over 670,000 employees, most of whom work decently and well every day; it is only a few, individual persons in the company who order or commit crimes. These persons must be identified and punished; a punishment of the company can indirectly also hit the innocent employees. On the other hand, the emissions scandal also shows how difficult it is to identify criminals in the company, especially when companies make it difficult to investigate. When the American prosecutor has reached his goal and proven the company guilty, the German public prosecutor starts painstaking detail work. In the USA, the same discussion took the other direction. Often only companies were charged there; the responsible managers got off with a handclap or a generous severance payment. After the banking crisis in 2008, a consensus developed to strengthen individual responsibility in criminal law, which led to a corresponding decree by the US Department of Justice in 2015 [Yates15]. The emissions scandal that became known thereafter was the first major application case of the Yates decree. In return, work was done in Germany on a corporate criminal law ("Law to Strengthen Integrity in Business") that also allows the punishment of companies. A sensible combination of personal criminal law with the new corporate criminal law might enable the rule of law to take action more successfully against economic crimes. However, after considerable resistance against such a law, its introduction seems unlikely in the near future.

Furthermore, the separation between civil law and criminal law in the USA is less sharp, so in the USA, compensations can also have a punitive character (punitive damages), i.e., exceed the amount of the damage, while in Europe compensations are smaller or equal to the damage. Especially in connection with the emissions scandal, the prosecution of crimes in the USA appeared consistent, while in Europe many alleged perpetrators have not yet been punished. In Germany, alleged main perpetrators are still at large. A comparative overview is given by [Rattalma17].

The most significant crime in connection with manipulations of emission values is fraud. A major difficulty in Germany compared to other European countries is the high requirement for fraud, in particular a financial loss is required, which only became beyond doubt after the discovery through the loss in value of diesel vehicles. A stock corporation like VW is obliged to publish events that are relevant for the operational result; the omission is punishable. The damage to the environment and the health of the general public has little legal significance in Germany in connection with the emissions scandal.

Since public prosecutors, unlike judges, are subject to instructions (§ 146 GVG) [GVG], it cannot be ruled out that political guidelines influence the investigations.

The following will provide an overview of ongoing and completed criminal proceedings in connection with the emissions scandal in Germany and the USA. Investigations are also being conducted in other EU countries, with the accused (often members of national subsidiaries such as Volkswagen Italia) and the allegations varying considerably.

5.6.1 Germany

In Germany, investigations have been/were conducted for (commercial and gang-related) fraud according to § 263 StGB, aiding and abetting fraud according to § 27 StGB, and for market manipulation according to [WpHG]. Due to the manipulations, vehicles were wrongly classified into the Euro 6 emission standard and temporarily exempted from motor vehicle tax, therefore investigations are also being conducted for tax evasion (§ 370 [AO]). An expert opinion [Führ16] sees an environmental crime enabled by collusion (here cooperation between perpetrators in companies and regulatory authorities, § 330d StGB).

The public prosecutor's offices responsible for the company headquarters are each investigating the companies and their partly unknown employees; for VW this is the public prosecutor's office in Braunschweig, for Opel the public prosecutor's office in Frankfurt, for Daimler and Porsche the public prosecutor's office in Stuttgart and for BMW and Audi the public prosecutor's offices I and II in Munich, in addition there are also investigations at suppliers.

There are criminal complaints for air pollution (§ 325 StGB), bodily harm (§ 223 StGB) and serious bodily harm (§ 226 StGB). No charges were brought against the perpetrators for this reason; in the case of bodily harm, the difficult burden of proof may also deter the responsible public prosecutors, as the health effects must be quantified and linked to the respective acts. The criminal processing of the emissions scandal is much more difficult at this point and must be considered as an open issue.

The criminal proceedings for market manipulation against the VW CEO Diess and the Chairman of the Supervisory Board Pötsch were discontinued in 2020 before the main proceedings at the LG Braunschweig against a payment of 4.5 million euros each [LG_BS20]; this is a common procedure according to § 153a StPO.

The proceedings of the public prosecutor's office Braunschweig against the former CEO of VW, Martin Winterkorn, were initially supposed to take place together with four other VW managers (the head of engine development and later development manager Heinz-Jakob Neußer, his predecessor in engine development Jens Hadler, the head of drive electronics Hanno Jelden and the team leader exhaust aftertreatment Thorsten D. [Bender21, Müssgens21]) (Az. 6 KLs 23/19), but were then separated due to a medical report and postponed indefinitely. There are also other criminal proceedings.

Of the four accused Audi managers Giovanni Pamio, Henning Lörch, Wolfgang Hatz and Rupert Stadler, three had to be held in pre-trial detention for several months due to the risk of tampering with evidence (destruction of evidence, influencing witnesses). Stadler was CEO of Audi AG at the time of his arrest; he is accused of fraud, indirect false certification (§ 271 StGB) and criminal advertising (§ 16 [UWG]). Lörch left the proceedings early with a fine after his confession and is therefore considered innocent. The other three defendants confessed late, their verdicts are expected to be delivered on 27.06.2023. The public prosecutor's office Munich II filed charges in August 2020 against the four former Audi managers Richard Bauder, Ulrich Hackenberg (former development board member), Stefan Knirsch and Bernd Martens for fraud, indirect false certification and criminal advertising.

When the head of engine development Jörg Kerner drove away during a search at Porsche, this was interpreted as an attempt to escape; he was taken into custody [Köster18]. The proceedings against the Porsche board member for research and development Michael Steiner were discontinued against payment, and a penalty order for fraud was issued against Kerner in December 2021 [Bender22].

The Stuttgart public prosecutor's office applied for penalty orders against three Daimler employees at the Böblingen district court in July 2021. The three employees, who are rather low in the hierarchy, received penalty orders for fraud with suspended sentences of less than one year.

5.6.2 USA

After the emission scandal became public, criminal proceedings against the perpetrators began swiftly in the USA.

James R. Liang was the first VW manager to be sentenced in 2017 to a prison term of 40 months without probation and a fine of 200,000 US$. Liang was significantly involved in the development of the EA189 engines for the US market. The relatively mild sentence by American standards is due to the fact that he played a minor role in the emission scandal and contributed to its resolution. He was handed over to German justice in 2019 and released early on probation there. His willingness to cooperate and his knowledge are likely to support the investigations.

Oliver Schmidt, responsible for VW's approvals in the USA until 2015, was arrested at the airport upon returning from a vacation in Florida in January 2017 and was sentenced the same year for conspiracy to commit fraud (most comparable to organized fraud according to § 263 (5) StGB in Germany, see also "attempted participation" according to § 30 StGB) and violation of environmental laws to a prison term of seven years without probation and a fine of 400,000 US$. He is now also in Germany and has been released early on probation.

Against the former VW CEO Martin Winterkorn, an arrest warrant was issued in the USA in 2018; he is accused of conspiracy to violate environmental laws and deception of authorities. In addition, he is being investigated for deceiving investors. If Winterkorn leaves Germany, he must expect his extradition to the USA (Germany does not extradite German citizens to the USA). He is likely to face a significantly longer prison sentence than Liang and Schmidt.

From the perspective of US justice, the former VW development board member Heinz-Jakob Neußer is still a significant perpetrator.

At the beginning of 2019, the four former Audi managers Bauder, Eiser, Knirsch, and Nagel were charged in absentia, outside of Germany, they face extradition to the USA; for Eiser, a vacation ended in Croatian extradition custody. Further indictments are to be expected [Stockb19].

5.7   Internal Investigations

The companies affected by the manipulation should have their own interest in clarifying the internal processes, on the one hand to identify the perpetrators within the company, on the other hand to be able to defend themselves more effectively externally. In the long term, internal investigations should document and eliminate weaknesses in the corporate culture.

The complexity and scope of the emission scandal may exceed the capabilities of internal auditing. Investigation methods include the review of documents and data carriers (in external investigations one would speak of a search) and conversations, which would be referred to as interrogations in external investigations.

VW has commissioned 2015 Jones Day, one of the largest law firms worldwide with 2400 lawyers, with the internal investigations, but it was involved in a scandal itself [WiWo16/2]. Other law firms, especially Gleiss Lutz, were commissioned with supplementary investigations. The internal investigations at VW are "completed" (results or consequences are not known); part of the internal investigation documents was seized in 2017 by the Munich II public prosecutor's office and could still be of importance in the ongoing criminal proceedings. A complaint against the seizure at the Federal Constitutional Court was not successful. As a result of the emission scandal, a position for integrity and law was created on the board, the first incumbent Christine Hohmann-Dennhardt remained from January 2016 to January 2017.

Other companies have also internally processed the emission scandal; however, the extent of internal investigations and also public interest were significantly lower than at VW.

Immediate consequences of the internal investigations for the perpetrators can be of labor law nature (warning, dismissal), civil law nature (claims for damages), and criminal law nature (criminal complaint).

Due to meticulously documented development processes, it is easy to find out which programmers and engineers have been involved in defeat devices. Although the individual guilt of individual developers is considered to be low, their statements can point the way to the responsible decision-makers. These statements are all the more important because requests for criminal offenses are usually not documented in writing.

5.8 TTIP

The Transatlantic Trade and Investment Partnership TTIP was a treaty negotiated between Europe and the USA from 2013 to 2016, initially under secrecy even from constitutional organs of the participating states. It was supported by the German automotive industry. A core of TTIP was, in addition to tariff relief and sensible standardization of technical standards, the reduction of non-monetary trade barriers. These can include, for example, environmental protection or consumer protection requirements of individual states that have a detrimental effect on the economic interests of global economic actors. TTIP envisaged a privately organized parallel justice to the rule of law, the companies involved in the emission scandal could possibly have sued the USA for their environmental requirements via these arbitration courts. The mere prospect of a lawsuit could have restricted democratically legitimized action.

5.9 Possible Improvements

5.9.1 Formulation of Legal Requirements

The unclear formulation of Art. 5 (2) a), Regulation 715/2007 has created a broad gray area between legality and illegality, which many manufacturers exploit to the limit and sometimes beyond.

A clear formulation primarily includes an unavoidable definition by the legislator of the temperature range in which exhaust emission limits must be met without exception. This may require independent investigations to define realistic temperature limits for different engine classes. However, the execution of these investigations should not justify any longer delays than objectively necessary.

5.9.2 Test Procedures

As early as 2008, the German Federal Ministry of the Environment noticed that current test procedures are unsuitable, especially in the case of manipulated exhaust values. Due to political sensitivity, it was decided at the time to forget this knowledge [Becker16] (on whose instruction is not known). In the mean-time, measures have been taken with the new WLTC cycle and the Real Driving Emissions (Sect. 3.1). After their introduction, their effectiveness remains to be observed. After the controversial conformity factors (Sect. 3.1.3) were lowered, these measures seem to be proving effective.

5.9.3 Code Analysis

Various parties have suggested that, in addition to the testing procedures from Sect. 3.1, the source code of the engine and possibly the transmission control unit should also be examined as part of the type approval process. The immense amount of work required to analyze the code is a critical issue, as the supervisory authorities do not have the necessary personnel, who must also be highly qualified (programming knowledge, engine knowledge). Practically, the problem could be solved by not checking every source code, but only in randomly selected or suspected control units, but with corresponding meticulousness. Defeat devices already known in a similar form can also be detected algorithmically [Contag17].

An interesting proposal was made before the EMIS committee of inquiry of the EU Parliament [LangeEU], namely the publication of the control unit source code (Open Source). Such a proposal contradicts the extremely pronounced cul-ture in the automotive industry of keeping even trivialities secret. In fact, there are no legitimate protection interests against disclosing the code. Software source code is protected by copyright, and commercial use by third parties after publi-cation requires the consent of the rights holder. In practice, however, violations may be difficult to prove. The code can also be part of an invention, which can be protected by patent law (direct patenting of software is not possible and not necessary due to copyright protection). Since such commercial legal protection is guaranteed, it would have to be questioned whether further protection inter-ests (e.g., against tuning) exist and whether they enjoy the same status as the public interest in subsequent and preventive clarification. The code for theft pro-tection functions is not suitable for publication. Further proposals are described in [Mock15].

Programmers develop the control unit software in a high-level programming language called C (source code), which is an intermediate level between human thinking and the primitive commands that a microcomputer can execute (machine code). The translation of the C source code into executable machine code is done with a translator (compiler), supplemented by auxiliary programs [Rechenb06]. Increasingly, the C code is no longer created by hand, but is automated from graphical representations that are even more in line with human thinking. The difficulty with any source code analysis is to ensure that the source code provided is actually the source from which the executable machine code of the control unit was generated by compilation.

A possible approach is

- to generate an identification number (hash code [Rechenb06]) from the machine code,
- to deliver this along with it,
- then have the supervisory authority compile the submitted source code and derive a hash code from it, which should be identical to the supplied hash code.

Laypeople may be familiar with such hash codes like MD5 because they can be used to check the error-free nature of larger downloads from the internet. The disadvantage is that with a very low probability, different programs can produce the same hash code (risk of manipulation) and that the source code must be compiled with exactly the same development environment and settings as the software supplier (control unit supplier or others).

Machine code can also be made partially readable by disassembly, it is then called assembler code. Knowledge about shutdown functions has reached the public in this way. However, the hardware-related assembler code is difficult to read even for programmers, as the following short fragment (estimated to be less than one-twentieth of the total code) from the VW acoustic function demonstrates:

```
...
mov16 d4,d9
ld32.a a15,a9(off_801E7368 - _40_a9)
ne d15,d15,#0
insert d15,d10,d15,#2,#1
extr.u d10,d15,#0,#8
...
```

Since even the more readable source code is not easy to interpret, at least when a code analysis is carried out by supervisory authorities, the unabridged documentation of the software (also called function frame) should be provided. Then, submitting a shortened or falsified documentation should itself be a punishable offense.

The EU regulation [EU18/858] allows supervisory authorities to enforce the release of software. There is still a need for regulation regarding the release of software and documents to courts in civil proceedings, as the availability of such documents from criminal proceedings in Germany is often not given.

Since a network of control units operates in the vehicle, an analysis of the engine control unit code cannot reveal the complete system behavior in the vehicle, as the interactions with other control units beyond the communication interfaces are not comprehensible.

References

[AEUV] *Treaty on the Functioning of the European Union,* consolidated verstion, Official Journal of the European Union, C 202, 7 June 2016. https://eur-lex.europa.eu/legal-content/EN/TXT/PDF/?uri=OJ:C:2016:202:FULL. Accessed: 21. Febr. 2023

[AO] *Abgabenordnung in der Fassung der Bekanntmachung vom 1. Oktober 2002 (BGBl. I S. 3866; 2003 I S. 61), die zuletzt durch Artikel 4 des Gesetzes vom 20. Dezember 2022 (BGBl. I S. 2730) geändert worden ist.* https://www.gesetze-im-internet.de/ao_1977/BJNR006130976.html. Accessed: 15. Febr. 2023

[Becker16] Becker, S., Rosenbach, M., Traufetter, G.: *Der gelöschte Verdacht.* Spiegel **38** (2016), 76–77

[Bender20] Bender, R., Keuchel, J., Votsmeier, V.: *Diesel-Richter gegen Diesel-Richter.* Handelsblatt **7** (2020), 16–17

[Bender21] Bender, R., Votsmeier, V.: *VW-Betrugsprozess startet heute: Vier Angeklagte und ein leerer Platz,* Handelsblatt, 16.09.2021, online: https://www.handelsblatt.com/unternehmen/industrie/dieselskandal-vw-betrugsprozess-startet-heute-vier-angeklagte-und-ein-leerer-platz/27614176.html. Accessed: 21. Febr. 2023

[Bender22] Bender, R.: *Glimpfliches Ende für Porsche-Manager: Staatsanwaltschaft erlässt Strafbefehl und Geldauflagen.* Handelsblatt, 11.04.2022. https://www.handelsblatt.com/unternehmen/industrie/dieselskandal-glimpfliches-ende-fuer-porsche-manager-staatsanwaltschaft-erlaesst-strafbefehl-und-geldauflagen/28229288.html. Accessed: 15. Febr. 2023

[BGB] *Bürgerliches Gesetzbuch in der Fassung der Bekanntmachung vom 2. Januar 2002 (BGBl. I S. 42, 2909; 2003 I S. 738), das zuletzt durch Artikel 6 des Gesetzes vom 7. November 2022 (BGBl. I S. 1982) geändert worden ist.* https://www.gesetze-im-internet.de/bgb. Accessed: 15. Febr. 2023

[BGH252/19] BGH: Judgement VI ZR 252/19 (25.05.2020). https://www.bundesgerichtshof.de/SharedDocs/Pressemitteilungen/DE/2020/2020063.html. Accessed: 15. Febr. 2023

[BGH397/19] BGH: Judgement VI ZR 397/19 (30.07.2020). https://www.bundesger ichtshof.de/SharedDocs/Pressemitteilungen/DE/2020/2020100.html. Accessed: 15. Febr. 2023

[BGH418/21] BGH: Judgement VIa ZR 418/21 (13.06.2022). https://www.bundes gerichtshof.de/SharedDocs/Pressemitteilungen/DE/2022/2022091.html. Accessed: 15. Febr. 2023

[BImSchV10] *Zehnte Verordnung zur Durchführung des Bundes-Immissionsschutzgesetzes*, Verordnung über die Beschaffenheit und die Auszeichnung der Qualitäten von Kraft- und Brennstoffen vom 8. Dezember 2010 (BGBl. I S. 1849), die zuletzt durch Artikel 1 der Verordnung vom 13. Dezember 2019 (BGBl. I S. 2739) geändert worden ist. https://www.gesetze-im-internet.de/bimschv_10_2010. Accessed: 21. Febr. 2023

[BImSchV13] *Dreizehnte Verordnung zur Durchführung des Bundes-Immissionsschutzgesetzes*, Verordnung über Großfeuerungs-, Gasturbinen- und Verbrennungsmotoranlagen vom 6. Juli 2021 (BGBl. I S. 2514). https://www.gesetze-im-internet.de/bimschv_13_2021. Accessed: 21. Febr. 2023

[BImSchV28] *Achtundzwanzigste Verordnung zur Durchführung des Bundes-Immissionsschutzgesetzes*, Verordnung über Emissionsgrenzwerte für Verbrennungsmotoren vom 21. Juli 2021 (BGBl. I S. 3125). https://www.gesetze-im-internet.de/bimschv_28_2021. Accessed: 21. Febr. 2023

[BImSchV35] *Fünfunddreißigste Verordnung zur Durchführung des Bundes-Immissionsschutzgesetzes*, Verordnung zur Kennzeichnung der Kraftfahrzeuge mit geringem Beitrag zur Schadstoffbelastung vom 10. Oktober 2006 (BGBl. I S. 2218), die zuletzt durch Artikel 85 der Verordnung vom 31. August 2015 (BGBl. I S. 1474) geändert worden ist. http://www.gesetze-im-internet.de/bimschv_35. Accessed: 15. Febr. 2023

[BImSchV39] *Neununddreißigste Verordnung zur Durchführung des Bundes-Immissionsschutzgesetzes*, Verordnung über Luftqualitätsstandards und Emissionshöchstmengen vom 2. August 2010 (BGBl. I S. 1065), die zuletzt durch Artikel 112 der Verordnung vom 19. Juni 2020 (BGBl. I S. 1328) geändert worden ist. http://www.gesetze-im-internet.de/bimschv_39. Accessed: 21. Febr. 2023

[BImSchV44] *Vierundvierzigste Verordnung zur Durchführung des Bundes-Immissionsschutzgesetzes*, Verordnung über mittelgroße Feuerungs-, Gasturbinen- und Verbrennungsmotoranlagen vom 13. Juni 2019 (BGBl. I S. 804), die zuletzt durch Artikel 1 der Verordnung vom 12. Oktober 2022 (BGBl. I S. 1801) geändert worden ist. https://www.gesetze-im-internet. de/bimschv_44. Accessed: 15. Febr. 2023

[BJA] Bundesjustizamt (German Federal Office of Justice). https://www.bundes justizamt.de/DE/Themen/Verbraucherrechte/Musterfeststellungsklagen/ Musterfeststellungsklagen_node.html. Accessed: 15. Febr. 2023

[BMVI16] Federal Ministry for Traffic and Digital Infrstructure: *Bericht der Unter-suchungskommission „Volkswagen"* (2016). https://www.kba.de/DE/The men/Marktueberwachung/Abgasthematik/erster_ber_uk_vw_nox.pdf;jse ssionid=472871D156856711E618BD0BA98DE6D0.live11292?__blob= publicationFile&v=2. Accessed: 15. Febr. 2023

[BorgWarner22] BorgWarner, Worldwide Emission Standards 2022/2023 Passenger Cars & Light Duty Vehicles und 2021/2022 Heavy Duty & Off-Highway Vehicles. https://www.borgwarner.com/technologies/emissions-standards. Accessed: 15. Febr. 2023,

[boschvw] Bosch Settlement Claims Administrator. https://boschvwsettlement.com. Accessed: 15. Febr. 2023

[Bundestag16] Scientific Services of the German Parliament, *Abschalteinrichtungen in Personenkraftwagen, Zur Reichweite des Verbots nach der Verordnung (EG) Nr. 715/2007*, report WD 7 – 3000 – 031/16 (16.03.2016). https:// www.bundestag.de/blob/417458/a55f9af383df0cf6862384d0b5b83611/ wd-7-031-16-pdf-data.pdf. Accessed: 15. Febr. 2023

[Cabraser16] Cabraser, E.J.: Complaint. http://www.cand.uscourts.gov/filelibrary/1709/ Consolidated_Consumer_Class_Action_Complaint.pdf.   Accessed: 15. Febr. 2023

[Contag17] Contag, M., Li, G., Pawlowski, A., Domke, F., Levchenko, K., Holz, T., Savage, S.: *How they did it: an analysis of emission defeat devices in modern automobiles*, IEEE Symposium on Security & Privacy, 2017, San Jose, CA, USA, https://cseweb.ucsd.edu/~klevchen/diesel-sp17.pdf. Accessed: 11. Febr. 2023

[Dieselnet] https://www.dieselnet.com/standards/cycles. Accessed: 15. Febr 203

[dpa16] Deutsche Presse-Agentur, published by heise online, *Abgas-Skandal: Enge Abstimmung von KBA und Herstellern*, 11.11.2016. https://www. heise.de/autos/artikel/Abgas-Skandal-Enge-Abstimmung-zwischen-KBA-und-Herstellern-3463860.html. Accessed: 15. Febr. 2023

[ECE24] *Regulation No 24 of the Economic Commission for Europe of the United Nations (UN/ECE) – Uniform provisions concerning: I. The approval of compression ignition (C.I.) engines with regard to the emission of visible pollutants – II. The approval of motor vehicles with regard to the installation of C.I. engines of an approved type – III. The approval of motor vehicles equipped with C.I. engines with regard to the emission of visible pollutants by the engine – IV. The measurement of power of C.I. engines*, Rev. 2 (1986). https://www.unece.org/fileadmin/DAM/ trans/main/wp29/wp29regs/r024r2e.pdf + 8 Amendments (2001–2023). Accessed: 15. Febr. 2023

[ECE40] *Regulation No 40 of the Economic Commission for Europe of the United Nations (UN/ECE) – Uniform provisions concerning the approval of motor cycles equipped with a positive-ignition engine with regard to the emission of gaseous pollutants by the engine* (1979). https://www.unece. org/fileadmin/DAM/trans/main/wp29/wp29regs/r040e.pdf + 4 corrections (1979–1996) und 2 amendments (1988–2007). Accessed: 15. Febr. 2023

[ECE47] *Regulation No 47 of the Economic Commission for Europe of the United Nations (UN/ECE) – Uniform provisions concerning the approval of mopeds equipped with a positive-ignition engine with regard to the emission of gaseous pollutants by the engine* (1981). https://www.unece.

org/fileadmin/DAM/trans/main/wp29/wp29regs/r047e.pdf + amendment (2007). Accessed: 15. Febr. 2023

[ECE49] *Regulation No 49 of the Economic Commission for Europe of the United Nations (UN/ECE) – Uniform provisions concerning the measures to be taken against the emission of gaseous and particulate pollutants from compression ignition engines and positive ignition engines for use in vehicles*, Rev. 8 (2023). https://www.unece.org/fileadmin/DAM/trans/main/wp29/wp29regs/2013/R049r6e.pdf + amendment (2023). Accessed: 15. Febr. 2023

[ECE83] *Regulation No 83 of the Economic Commission for Europe of the United Nations (UN/ECE) – Uniform provisions concerning the approval of vehicles with regard to the emission of pollutants according to engine fuel requirements*, Rev. 5 (2015). https://www.unece.org/fileadmin/DAM/trans/main/wp29/wp29regs/R083r5e.pdf + 14 amendments (2016–2022). Accessed: 15. Febr. 2023

[ECE84] *Regulation No 84 of the Economic Commission for Europe of the United Nations (UN/ECE) – Uniform provisions concerning the approval of passenger cars equipped with an internal combustion engine with regard to the measurement of fuel consumption* (1990). https://www.unece.org/fileadmin/DAM/trans/main/wp29/wp29regs/R084e.pdf. Accessed: 15. Febr. 2023

[ECE85] *Regulation No 85 of the Economic Commission for Europe of the United Nations (UN/ECE) – Uniform provisions concerning the approval of internal combustion engines or electric drive trains intended for the propulsion of motor vehicles of categories M and N with regard to the measurement of net power and the maximum 30 minutes power of electric drive trains*, Rev. 1 (2013). https://www.unece.org/fileadmin/DAM/trans/main/wp29/wp29regs/2015/R085r1e.pdf + correction (2016) + 4 amendments (2016–2020). Accessed: 15. Febr. 2023

[ECE101] *Regulation No 101 of the Economic Commission for Europe of the United Nations (UN/ECE) – Uniform provisions concerning the approval of passenger cars powered by an internal combustion engine only, or powered by a hybrid electric power train with regard to the measurement of the emission of carbon dioxide and fuel consumption and/or the measurement of electric energy consumption and electric range, and of categories M_1 and N_1 vehicles powered by an electric power train only with regard to the measurement of electric energy consumption and electric range*, Rev. 3 (2013). https://www.unece.org/fileadmin/DAM/trans/main/wp29/wp29regs/2015/R101r3e.pdf + 10 amendments (2013–2022). Accessed: 15. Febr. 2023

[ECE_GTR2] *Measurement procedure for two-wheeled motorcycles equipped with a positive or compression ignition engine with regard to the emission of gaseous pollutants, CO2 emissions and fuel consumption* (ECE/TRANS/180/Add.2). https://unece.org/fileadmin/DAM/trans/main/wp29/wp29wgs/wp29gen/wp29registry/ECE-TRANS-180a2e.pdf + errata, corrigenda, appendices and amendments. Accessed: 30. Jan. 2023

[EPA]   EPA, Clean Air Act, Title II – *Emission Standards for Moving Sources*, Washington D. C. https://www.epa.gov/clean-air-act-overview/title-ii-emission-standards-moving-sources. Accessed: 15. Febr. 2023

[EU01/80]   *Directive 2001/80/EG of the European Parliament and the Council of 23 October 2001 on the limitation of emissions of certain pollutants into the air from large combustion plants.* https://eur-lex.europa.eu/LexUriServ/LexUriServ.do?uri=CONSLEG:2001L0080:20070101:EN:PDF. Accessed: 15. Febr. 2023

[EU01/81]   *Directive 2001/81/EG of the European Parliament of 23 October 2001 on national emission ceilings for certain atmospheric pollutants.* https://eur-lex.europa.eu/LexUriServ/LexUriServ.do?uri=OJ:L:2001:309:0022:0030:EN:PDF. Accessed: 15. Febr. 2023

[EU02/51]   *Directive 2002/51/EC of the European Parliament and the Council of 19 July 2002 on the reduction of the level of pollutant emissions from two- and three-wheel motor vehicles and amending Directive 97/24/EC.* https://eur-lex.europa.eu/legal-content/EN/TXT/PDF/?uri=CELEX:32002L0051. Accessed: 15. Febr. 2023

[EU03/77]   *Commission Directive 2003/77/EC of 11 August 2003 amending Directives 97/24/EC and 2002/24/EC of the European Parliament and of the Council relating to the type-approval of two- or three-wheel motor vehicles.* https://eur-lex.europa.eu/legal-content/EN/TXT/PDF/?uri=CELEX:32003L0077. Accessed: 15. Febr. 2023

[EU04/107]   *Directive 2004/107/EC of the European Parliament and the Council of 15 December 2004 relating to arsenic, cadmium, mercury, nickel and polycyclic aromatic hydrocarbons in ambient air.* https://eur-lex.europa.eu/LexUriServ/LexUriServ.do?uri=OJ:L:2005:023:0003:0016:EN:PDF. Accessed: 15. Febr. 2023

[EU05/30]   *Commission Directive 2005/30/EC of 22 April 2005 amending, for the purposes of their adaptation to technical progress, Directives 97/24/EC and 2002/24/EC of the European Parliament and of the Council, relating to the type-approval of two or three-wheel motor vehicles.* https://eur-lex.europa.eu/legal-content/EN/TXT/PDF/?uri=CELEX:32005L0030. Accessed: 15. Febr. 2023

[EU06/72]   *Commission Directive 2006/72/EC of 18 August 2006 amending for the purposes of adapting to technical progress Directive 97/24/EC of the European Parliament and of the Council on certain components and characteristics of two or three-wheel motor vehicles.* https://eur-lex.europa.eu/legal-content/EN/TXT/PDF/?uri=CELEX:32006L0072. Accessed: 15. Febr. 2023

[EU06/120]   *Commission Directive 2006/120/EG of 27 November 2006 correcting and amending Directive 2005/30/EC amending, for the purposes of their adaptation to technical progress, Directives 97/24/EC and 2002/24/EC of the European Parliament and of the Council, relating to the type-approval of two or three-wheel motor vehicles.* https://eur-lex.europa.eu/legal-content/EN/TXT/PDF/?uri=CELEX:32006L0120. Accessed: 15. Febr. 2023

[EU07/46]   *Directive 2007/46/EG of the European Parliament and the Council of 5 September 2007 establishing a framework for the approval of motor vehicles and their trailers, and of systems, components and separate technical units intended for such vehicles (Framework Directive).* https://eur-lex.eur opa.eu/legal-content/EN/TXT/PDF/?uri=CELEX:32007L0046. Accessed: 15. Febr. 2023

[EU07/715]   *Regulation (EC) No 715/2007 of the European Parliament and the Council of 20 June 2007 on type approval of motor vehicles with respect to emissions from light passenger and commercial vehicles (Euro 5 and Euro 6) and on access to vehicle repair and maintenance information.* https://eur-lex.eur opa.eu/legal-content/EN/TXT/PDF/?uri=CELEX:32007R0715. Accessed: 15. Febr. 2023

[EU08/50]   *Directive 2008/50/EC of the European Parliament and of the Council of 21 May 2008 on ambient air quality and cleaner air for Europe.* https:// eur-lex.europa.eu/legal-content/EN/TXT/?uri=CELEX%3A32008L0050, consolidated version after correction and change by directive 2015/1480: https://eur-lex.europa.eu/legal-content/EN/TXT/?uri=CELEX%3A0200 8L0050-20150918. Accessed: 15. Febr. 2023

[EU08/692]   *Commission Regulation (EC) No 692/2008 of 18 July 2008 implementing and amending Regulation (EC) No 715/2007 of the European Parliament and of the Council on type-approval of motor vehicles with respect to emissions from light passenger and commercial vehicles (Euro 5 and Euro 6) and on access to vehicle repair and maintenance information.* https://eur-lex.eur opa.eu/legal-content/EN/TXT/PDF/?uri=CELEX:32008R0692. Accessed: 15. Febr. 2023

[EU09/30]   *Directive 2009/30/EG of the European Parliament of 23 April 2009 amending Directive 98/70/EC as regards the specification of petrol, diesel and gas-oil and introducing a mechanism to monitor and reduce greenhouse gas emissions and amending Council Directive 1999/32/EC as regards the specification of fuel used by inland waterway vessels and repealing Directive 93/12/EEC.* https://eur-lex.europa.eu/legal-content/EN/TXT/ PDF/?uri=CELEX:32009L0030. Accessed: 15. Febr. 2023

[EU09/108]   *Commission Directive 2009/108/EC of 17 August 2009 amending, for the purposes of adapting it to technical progress, Directive 97/24/EC of the European Parliament and of the Council on certain components and characteristics of two or three-wheel motor vehicles.* https://eur-lex.europa. eu/legal-content/EN/TXT/PDF/?uri=CELEX:32009L0108. Accessed: 15. Febr. 2023

[EU09/595]   *Regulation (EC) No 595/2009 of the European Parliament and the Council of 18 June 2009 on type-approval of motor vehicles and engines with respect to emissions from heavy duty vehicles (Euro VI) and on access to vehicle repair and maintenance information and amending Regulation (EC) No 715/2007 and Directive 2007/46/EC and repealing Directives 80/1269/EEC, 2005/55/EC and 2005/78/EC.* https://eur-lex.eur opa.eu/legal-content/EN/TXT/PDF/?uri=CELEX:32009R0595. Accessed: 15. Febr. 2023

[EU10/75] *Directive 2010/75/EU of the European Parliament and the Council of 24 November 2010 on industrial emissions (integrated pollution prevention and control).* https://eur-lex.europa.eu/legal-content/EN/TXT/PDF/?uri=CELEX:32010L0075. Accessed: 15. Febr. 2023

[EU11/566] *Commission Regulation (EU) N0 566/2011 of 8. June 2011 amending Regulation (EC) No 715/2007 of the European Parliament and of the Council and Commission Regulation (EC) No 692/2008 as regards access to vehicle repair and maintenance information.* https://eur-lex.europa.eu/legal-content/EN/TXT/PDF/?uri=CELEX:32011R0566. Accessed: 15. Febr. 2023

[EU11/582] *Commission Regulation (EU) No. 582/2011 of 25 May 2011 implementing and amending Regulation (EC) No 595/2009 of the European Parliament and of the Council with respect to emissions from heavy duty vehicles (Euro VI) and amending Annexes I and III to Directive 2007/46/EC of the European Parliament and of the Council.* https://eur-lex.europa.eu/legal-content/EN/TXT/PDF/?uri=CELEX:32011R0582, with changes to 2019: https://eur-lex.europa.eu/legal-content/EN/TXT/PDF/?uri=CELEX:02011R0582-20191215. Accessed: 15. Febr. 2023

[EU12/459] *Commission Regulation (EU) No 459/2012 of 29 May 2012 amending Regulation (EC) No 715/2007 of the European Parliament and of the Council and Commission Regulation (EC) No 692/2008 as regards emissions from light passenger and commercial vehicles (Euro 6).* https://eur-lex.europa.eu/legal-content/EN/TXT/PDF/?uri=CELEX:32012R0459. Accessed: 15. Febr. 2023

[EU13/60] *Commission Directive 2013/60/EU of 27 November 2013 amending for the purposes of adapting to technical progress, Directive 97/24/EC of the European Parliament and of the Council on certain components and characteristics of two or three-wheel motor vehicles, Directive 2002/24/EC of the European Parliament and of the Council relating to the type-approval of two or three-wheel motor vehicles and Directive 2009/67/EC of the European Parliament and of the Council on the installation of lighting and light-signalling devices on two- or three-wheel motor vehicles.* https://eur-lex.europa.eu/legal-content/EN/TXT/PDF/?uri=CELEX:32013L0060. Accessed: 15. Febr. 2023

[EU13/168] *Regulation (EU) No 168/2013 of the European Parliament and the Council of 15 January 2013 on the approval and market surveillance of two- or three-wheel vehicles and quadricycles.* https://eur-lex.europa.eu/legal-content/EN/TXT/PDF/?uri=CELEX:32013R0168. Accessed: 15. Febr. 2023

[EU14/134] *Commission Delegated Regulation (EU) No. 134/2014 of 16 December 2013 supplementing Regulation (EU) No 168/2013 of the European Parliament and of the Council with regard to environmental and propulsion unit performance requirements and amending Annex V thereof.* https://eur-lex.europa.eu/legal-content/EN/TXT/PDF/?uri=CELEX:32014R0134. Accessed: 15. Febr. 2023

[EU14/136] *Commission Regulation (EU) No 136/2014 of 11 February 2014 amending Directive 2007/46/EC of the European Parliament and of the Council, Commission Regulation (EC) No 692/2008 as regards emissions from light passenger and commercial vehicles (Euro 5 and Euro 6) and Commission Regulation (EU) No 582/2011 as regards emissions fromheavy duty vehicles (Euro VI).* https://eur-lex.europa.eu/legal-content/EN/TXT/PDF/?uri= CELEX:32014R0136. Accessed: 15. Febr. 2023

[EU16/1628] *Regulation (EU) 2016/1628 of the European Parliament and the Council of 14 September 2016 on requirements relating to gaseous and particulate pollutant emission limits and type-approval for internal combustion engines for non-road mobile machinery, amending Regulations (EU) No 1024/2012 and (EU) No 167/2013, and amending and repealing Directive 97/68/EC.* https://eur-lex.europa.eu/legal-content/EN/TXT/PDF/?uri= CELEX:32016R1628. Accessed: 15. Febr. 2023

[EU16/1824] *Commission Delegated Regulation (EU) 2016/1824 of 14 July 2016 amending Delegated Regulation (EU) No 3/2014, Delegated Regulation (EU) No 44/2014 and Delegated Regulation (EU) No 134/2014 with regard, respectively, to vehicle functional safety requirements, to vehicle construction and general requirements and to environmental and propulsion unit performance requirements.* https://eur-lex.europa.eu/legal-content/EN/TXT/PDF/?uri=CELEX:32016R1824. Accessed: 15. Febr. 2023

[EU17/1151] *Commission Regulation (EU) 2017/1151 of 1 June 2017 supplementing Regulation (EC) No 715/2007 of the European Parliament and of the Council on type-approval of motor vehicles with respect to emissions from light passenger and commercial vehicles (Euro 5 and Euro 6) and on access to vehicle repair and maintenance information, amending Directive 2007/46/ EC of the European Parliament and of the Council, Commission Regulation (EC) No 692/2008 and Commission Regulation (EU) No 1230/2012 and repealing Commission Regulation (EC) No 692/2008.* https://eur-lex.eur opa.eu/legal-content/EN/TXT/PDF/?uri=CELEX:32017R1151. Accessed: 15. Febr. 2023

[EU17/1154] *Commission Regulation (EU) 2017/1154 of 7 June 2017 amending Regulation (EU) 2017/1151 supplementing Regulation (EC) No 715/2007 of the European Parliament and of the Council on type-approval of motor vehicles with respect to emissions from light passenger and commercial vehicles (Euro 5 and Euro 6) and on access to vehicle repair and maintenance information, amending Directive 2007/46/EC of the European Parliament and of the Council, Commission Regulation (EC) No 692/2008 and Commission Regulation (EU) No 1230/2012 and repealing Regulation (EC) No 692/2008 and Directive 2007/46/EC of the European Parliament and of the Council as regards real-driving emissions from light passenger and commercial vehicles (Euro 6).* https://eur-lex.europa.eu/legal-content/EN/TXT/ PDF/?uri=CELEX:32017R1154. Accessed: 15. Febr. 2023

[EU18/295] *Commission Delegated Regulation (EU) 2018/295 of 15 December 2017 amending Delegated Regulation (EU) No 44/2014, as regards vehicle construction and general requirements, and Delegated Regulation (EU) No 134/2014, as regards environmental and propulsion unit performance requirements for the approval of two- or three-wheel vehicles and quadricycles.* https://eur-lex.europa.eu/legal-content/EN/TXT/PDF/?

uri=CELEX:32018R0295. Accessed: 15. Febr. 2023

[EU18/858]   *Regulation (EU) 2018/858 of the European Council and the Parliament of 30 May 2018 on the approval and market surveillance of motor vehicles and their trailers, and of systems, components and separate technical units intended for such vehicles, amending Regulations (EC) No 715/2007 and (EC) No 595/2009 and repealing Directive 2007/46/EC.* https://eur-lex.eur opa.eu/legal-content/EN/TXT/PDF/?uri=CELEX:32018R0858. Accessed: 15. Febr. 2023

[EU18/1832]   *Commission Regulation (EU) 2018/1832 of 5 November 2018 amending Directive 2007/46/EC of the European Parliament and of the Council, Commission Regulation (EC) No 692/2008 and Commission Regulation (EU) 2017/1151 for the purpose of improving the emission type approval tests and procedures for light passenger and commercial vehicles, including those for in-service conformity and real-driving emissions and introducing devices for monitoring the consumption of fuel and electric energy.* https://eur-lex.europa.eu/legal-content/ENTXT/PDF/?uri= CELEX:32018R1832. Accessed: 15. Febr. 2023

[EU20/1694]   *Regulation (EU) 2020/1694 f the European Parliament and the Council of 11 November 2020 amending Regulation (EU) No 168/2013 as regards specific measures on L-category end-of-series vehicles in response to the COVID-19 pandemic.* https://eur-lex.europa.eu/legal-content/EN/TXT/ PDF/?uri=CELEX:32020R1694. Accessed: 30. Jan. 2023

[EU20/1828]   *Directive (EU) 2020/1828 of the European Parliament and the Council of 25 November 2020 on representative actions for the protection of the collective interests of consumers and repealing Directive 2009/22/EC.* https://eur-lex.europa.eu/legal-content/EN/TXT/PDF/? uri=CELEX:32020L1828. Accessed: 15. Febr. 2023

[EU21/1445]   *Commission Delegated Regulation (EU) 2021/1445 of 23 June 2021 amending Annexes II and VII to Regulation (EU) 2018/858 of the European Parliament and of the Council.* https://eur-lex.europa.eu/legal-content/EN/ TXT/?uri=CELEX:32021R1445. Accessed: 14. Febr. 2023

[EU23/443]   *Commission Regulation (EU) 2023/443 of 8 February 2023 amending Regulation (EU) 2017/1151 as regards the emission type approval procedures for light passenger and commercial vehicles.* https://eur-lex.europa. eu/legal-content/EN/TXT/PDF/?uri=CELEX:32023R0443. Accessed: 15. June 2023

[EU23]   European Union: *Non-transposition of EU legislation: Commission takes action to ensure complete and timely transposition of EU directives,* Press Release, 27.01.2023, https://ec.europa.eu/commission/presscorner/ detail/en/inf_23_262. Accessed: 21. Febr. 2023

[EU91/441]   *DirectiveCouncil 91/441/EEC of 26 June 1991 amending Directive 70/ 220/EEC on the approximation of the laws of the Member States relating to measures to be taken against air pollution by emissions from motor vehicles.* https://eur-lex.europa.eu/legal-content/EN/ TXT/PDF/?uri=CELEX:31991L0441. Accessed: 15. Febr. 2023

[EU97/24]   *Directive 97/24/EC of the European Parliament and the Council of 17 June 1997 on certain components and characteristics of two or three-wheel motor vehicles.* https://eur-lex.europa.eu/legal-content/EN/TXT/PDF/?uri=CELEX:01997L0024-20090907. Accessed: 15. Febr. 2023

[EuGH20]   EuGH, Case C-693/18, Judgement of 17.12.2020

[EuGH22]   EuGH, Case C-128/20, Judgement of 10.07.2022

[Flaig21]   Flaig, I.: *Das haben Daimler, Bosch & Co. schon bezahlt.* Stuttgarter Nachrichten, 07.07.2021 https://www.stuttgarter-nachrichten.de/inhalt.dieselskandal-das-haben-daimler-bosch-co-schon-bezahlt.37fcbbc2-793e-4d9f-bb32-0234dd862c86.html. Accessed: 14. Febr. 2023

[Führ16]   Führ, M.: *Gutachterliche Stellungnahme für den Deutschen Bundestag – 5. Untersuchungsausschuss der 18. Wahlperiode, fortgeschriebene Fassung,* 19.11.2016. https://www.bundestag.de/resource/blob/481344/c6f582c8598c9d6b62fcfb2acd012462/stellungnahme-prof--dr--fuehr--sv-4--data.pdf. Accessed: 15. Febr. 2023

[GVG]   *Gerichtsverfassungsgesetz in der Fassung der Bekanntmachung vom 9. Mai 1975 (BGBl. I S. 1077), das zuletzt durch Artikel 5 des Gesetzes vom 19. Dezember 2022 (BGBl. I S. 2606) geändert worden ist.* https://www.gesetze-im-internet.de/gvg. Accessed: 15. Febr. 2023

[HGB]   *Handelsgesetzbuch in der im Bundesgesetzblatt Teil III, Gliederungsnummer 4100–1, veröffentlichten bereinigten Fassung, das zuletzt durch Artikel 1 des Gesetzes vom 15. Juli 2022 (BGBl. I S. 1146) geändert worden ist.* https://www.gesetze-im-internet.de/hgb/BJNR002190897.html. Accessed: 14. Febr. 2023

[Heese]   Heese, M.: Projekt Dieselskandal: Herstellerhaftung. https://go.ur.de/dieselskandal. Accessed: 15. Febr. 2022

[Köster18]   Köster, K.: Razzia: *Porsche-Manager fuhr nach Hause – Verhaftung.* Stuttgarter Nachrichten 20.04.2018. https://www.stuttgarter-nachrichten.de/inhalt.durchsuchungen-wegen-diesel-razzia-porsche-manager-fuhr-nach-hause-verhaftung.2292b989-c197-49c2-84e5-3c8be6230764.html. Accessed: 15. Febr. 2023

[LangeEU]   Lange, D.: Statement to committee of inquiry EMIS of the European Parliament (16.06.2016). https://www.europarl.europa.eu/committees/en/emis/events-hearings.html?id=20160616CHE00302. Accessed: 15. Febr. 2023

[LG_BS16]   LG Braunschweig, 4 O 202/16, Judgement of 12.10.2016

[LG_BS20]   LG Braunschweig: Press Release of 20.05.2020. https://landgericht-braunschweig.niedersachsen.de/aktuelles/verdacht-der-marktmanipulation-strafverfahren-gegen-potsch-und-dr-diess-im-zwischenverfahren-eingestellt-188552.html. Accessed: 15. Febr. 2023

[LG_F19]   LG Frankfurt/Main, 2–19 O 279/19, Judgement of 07.08.2020

[LG_IN18]   LG Ingolstadt, 41 O 1745/18, Decision of 07.08.2020

[LG_M16]   LG München I, 23 O 23033/15, Judgement of 14.04.2016

[LG_KR16]   LG Krefeld, 2 O 72/16, Judgement of 14.09.2016

[Lübben21]   Lübben, T.: *Abgas-Skandal: Opel kommt mit Geldbuße davon.* Hessen-schau, 19.10.2021 https://www.hessenschau.de/wirtschaft/65-millionen-euro-strafe-opel-kommt-im-abgas-skandal-mit-geldbusse-davon,opel-abgas-skandal-100.html. Accessed: 14. Febr. 2023

[Mock15]   Mock, P., German, J.: *The future of vehicle emissions testing and compli-ance. How to align regulatory requirements, customer expectations, and environmental performance in the European Union,* ICCT (2015). https://www.theicct.org/sites/default/files/publications/ICCT_future-vehicle-tes ting_20151123.pdf. Accessed: 15. Febr. 2023

[Mussler21]   Mussler, W., Germis, C., Peitsmeier, H.: *Deutsche Autobauer müssen 875 Millionen Euro zahlen.* Frankfurter Allgemeine, 08.07.2021, online: https://www.faz.net/aktuell/wirtschaft/deutsche-autobauer-zahlen-in-eu-kartellverfahren-875-millionen-17427779.html. Accessed: 21. Febr. 2023

[Müssgens21]   Müssgens, Chr.: *Erste Runde ohne Winterkorn beendet.* Frankfurter Allgemeine, 30.09.2021, online: https://www.faz.net/aktuell/wirtschaft/unternehmen/top-manager-gegen-tueftler-17563553.html?pageIndex_2. Accessed: 15. Febr. 2023

[OLG_CE16]   OLG Celle, 7 W 26/16, Decision of 30.06.2016

[OLG_M19]   OLG München, 3 U 7392/19, Decision of 10.03.2020

[Rattalma17]   Frigessi di Rattalma, M. (Hrsg.): *The Dieselgate; A legal perspective.* Springer, Cham (2017)

[Rechenb06]   Rechenberg, P., Pomberger, G.: *Informatik-Handbuch.* Hanser, München (2006)

[Rogert16]   Rogert, M.: *Lieber in Luxemburg: Der Zulassungs-Tourismus deutscher Autohersteller.* Focus (27.05.2016). https://www.focus.de/auto/news/abgas-skandal/abgas-skandal-zulassungstourismus-nur-ein-deutscher-her steller-laesst-alle-autos-in-deutschland-zu_id_5571052.html. Accessed: 15. Febr. 2023

[StGB]   *Strafgesetzbuch in der Fassung der Bekanntmachung vom 13. November 1998 (BGBl. I S. 3322), das zuletzt durch Artikel 4 des Gesetzes vom 4. Dezember 2022 (BGBl. I S. 2146) geändert worden ist.* https://www.ges etze-im-internet.de/stgb. Accessed: 15. Febr. 2023

[Stockb19]   Stockburger, M., Schnell, A.: *Audi: US-Justiz erhebt Anklage gegen Knirsch und Bauder.* Heilbronner Stimme (2019). https://www.stimme. de/heilbronn/wirtschaft/2018/Audi-US-Justiz-erhebt-Anklage-gegen-Kni rsch-und-Bauder;art140955,4142293. Accessed: 15. Febr. 2023

[StPO]   *Strafprozeßordnung in der Fassung der Bekanntmachung vom 7. April 1987 (BGBl. I S. 1074, 1319), die zuletzt durch Artikel 2 des Gesetzes vom 25. März 2022 (BGBl. I S. 571) geändert worden ist.* https://www.gesetze-im-internet.de/stpo/index.html. Accessed: 15. Febr. 2023

[TALuft]   *Neufassung der Ersten Allgemeinen Verwaltungsvorschrift zum Bundes-Immissionsschutzgesetz (Technische Anleitung zur Reinhaltung der Luft – TA Luft) vom 18. August 2021.* https://www.verwaltungsvorschriften-im-internet.de/bsvwvbund_18082021_IGI25025005.htm. Accessed: 15. Febr. 2023

[TAFV] *Verordnung über technische Anforderungen an Transportmotorwagen und deren Anhänger (TAFV 1) vom 19. Juni 1995 (Stand am 1. Februar 2019).* https://www.fedlex.admin.ch/eli/cc/1995/4145_4145_4145/de. Accessed: 24. Jan. 2023

[TRGS554] *Technische Regeln für Gefahrstoffe – Abgase von Dieselmotoren (kurz TRGS 554) vom Januar 2019 (GMBl 2019 pp. 88–104 v. 18.03.2019)*

[UWG] *Gesetz gegen den unlauteren Wettbewerb in der Fassung der Bekanntmachung vom 3. März 2010 (BGBl. I S. 254), das zuletzt durch Artikel 20 des Gesetzes vom 24. Juni 2022 (BGBl. I S. 959) geändert worden ist.* https://www.gesetze-im-internet.de/uwg_2004. Accessed: 15. Febr. 2023

[Warentest] Stiftung Warentest. Verbraucherfreundliche Gerichtsentscheidungen. https://www.test.de/Abgasskandal-4918330-5038098/. Accessed: 15. Febr. 2023

[WiWo16/2] Bergermann, M.: *Kanzlei hat Ärger mit eigener Affäre.* Wirtschaftswoche, 23.02.2016

[WpHG] *Wertpapierhandelsgesetz in der Fassung der Bekanntmachung vom 9. September 1998 (BGBl. I S. 2708), das zuletzt durch Artikel 10 des Gesetzes vom 19. Dezember 2022 (BGBl. I S. 2606) geändert worden ist.* https://www.gesetze-im-internet.de/wphg. Accessed: 15. Febr. 2023

[Yates15] Yates, S. Q., Deputy Attorney General: *Individual Accountability for Corporate Wrongdoing.* 09.09.2015. https://www.justice.gov/archives/dag/file/769036/download. Accessed: 15. Febr. 2023

[ZPO] *Zivilprozessordnung in der Fassung der Bekanntmachung vom 5. Dezember 2005 (BGBl. I S. 3202; 2006 I S. 431; 2007 I S. 1781), die zuletzt durch Artikel 2 des Gesetzes vom 7. November 2022 (BGBl. I S. 1982) geändert worden ist.* https://www.gesetze-im-internet.de/zpo. Accessed: 15. Febr. 2023

Politics 6

6.1 Origins of Emission Legislation

In the mid of the last century, many countries in Europe were severely damaged by the Second World War. Those who lived among ruins, stole coal from freight trains to keep their meager accommodations warm, and worried about getting enough to eat, were not inclined to give high priority to environmental protection. The individual goal was to stay warm and fed, linked to the societal goal of reviving a collapsed economy. This was successfully achieved, as Germany and many other European countries experienced strong economic growth, which continued, albeit at a slower pace, for a long time. Environmental protection initially had a low priority, especially as it was long considered contrary to economic growth, while today it is seen more as a growth factor, albeit not without restrictions. The severe smog disaster that cost many lives in London in 1952 brought the issue of environmental protection onto the political agenda in Great Britain. In other countries, solid fuels were replaced by cleaner-burning liquid and gaseous fuels. Emissions from heating stoves decreased, but at the same time, motorization, previously reserved for the wealthy, became a mass phenomenon. Road traffic became the biggest polluter, surpassing industry and domestic heating. This led to a change in public attitude since the 1970s. In 1985, the smog alarm in the Ruhr area, which unlike the London smog disaster was no longer predominantly caused by heating, raised public awareness again. A clean environment and thus health became part of the quality of life. It would go beyond the scope of this book to discuss whether the onset of prosperity saturation, the increasing environmental impact from road traffic, or increasing knowledge about the health hazards of pollutants led to this. Although Willy Brandt already addressed the blue sky

© The Author(s), under exclusive license to Springer Fachmedien Wiesbaden GmbH, part of Springer Nature 2025
K. Borgeest, *Manipulation of Exhaust Gas Values*,
https://doi.org/10.1007/978-3-658-45864-5_6

over the Ruhr area as a goal in the 1961 election campaign, for a long time politics did little to protect the environment, which led to the formation of parties with an environmental focus in Europe from 1980 onwards. Meanwhile, the issue has arrived in almost all parties with varying priorities, not only in Europe.

Another development was the goal of reducing fuel consumption, which technically is identical to reducing CO_2. The suspected contribution of CO_2 to the greenhouse effect was for a long time a specialist topic of a few scientists without publicly recognizable relevance. The still low costs were no incentive to reduce fuel consumption. A change in thinking began with the oil crisis of 1973. The oil-producing countries reduced production, and the oil price reached levels then considered high. Although production was only reduced by a few percent, the Sunday driving bans in Germany and Switzerland and other measures in other countries impressively conveyed to the population the finiteness of oil reserves. A consensus in the population and politics on the need to save fuel and heating oil developed, to which vehicle manufacturers responded with more fuel-efficient engines. The issue later lost importance compared to pollutant reduction and power increase, which was supposed to enable the increasingly heavy vehicles to maintain their usual drivability. However, the indications that the increasing CO_2 content of the atmosphere caused by humans contributes to the greenhouse effect have become more concrete, so that politics was expected to act. This made the reduction of fuel consumption important again. Current EU CO_2 targets are defined in [EU19/631] and in smaller regulations for their adjustment.

Parallel to these environmental policy developments, European integration took place in Western Europe, and after 1990 also in large parts of Eastern Europe. In 1951, initially six European countries, including Germany, founded a common market for coal and steel, the European Coal and Steel Community; in 1957, the same countries founded the European Economic Community, which was later joined by other countries. The Economic Community evolved into an increasingly political union, which later renamed itself the "European Community" and finally the "European Union" (EU) [EUV92]. Thus, a large part of the early, then still sparse, national environmental laws were replaced by European legislation. Especially in the area of air pollutants and vehicle emissions of interest here, national laws almost exclusively implement European law. This seems sensible, as competition in the EU is not distorted by different environmental standards and as state borders do not influence the spread of pollutants. Switzerland is not a member of the EU, but has essentially adopted European emission legislation since 1995. In some areas, Switzerland is more consistent, e.g. with particle filters for construction machinery [LRV-CH].

For the political process of exhaust emission legislation, the process of European institutions is therefore relevant, in which German or other national constitutional bodies do not directly participate, but to a limited extent indirectly. In the member states, including Germany, there is typically a structure consisting of a parliament (legislature) that is supposed to convert the will of the people into laws, and an executing government (executive) that participates in the drafting of laws to varying degrees, but can also act without parliament within a defined scope, as in Germany through the issuance of regulations. The most important German regulations in the context of this book are the regulations on the Federal Immission Control Act, the Road Traffic Regulations [StVO] and the Road Traffic Licensing Regulations [StVZO], which largely have the task of implementing European law. The European structures are similar but not identical, there is a European Parliament (comparable to the German Bundestag), a European Commission (comparable to the German Federal Government) and a Council in which the member states are represented (comparable to the German Bundesrat, in which the federal states are represented). A significant difference to many nation states lies in the so far lesser (although in recent years, particularly with the Treaty of Lisbon [EU07], increased) influence of the European Parliament as a representation of the people and a resulting stronger concentration of power on the European Commission. Since pollutants do not know borders, a treaty under international law was already concluded in 1979, far beyond the EU (including North America and the then Soviet Union), the Geneva Air Pollution Agreement.

6.2 Monitoring of Legal Regulations

Is it a coincidence that the emission scandal was discovered in the USA and is being clarified following hints given by the environmental organization ICCT to the Californian environmental agency CARB in May 2014 after investigations at West Virginia University? To answer this question, it is worth looking at how it is monitored in Europe whether vehicle manufacturers comply with legal obligations. The European legislator has indeed defined exhaust emission limits and the technical procedures for their determination, but the organizational framework and in particular the sanctioning of violations are the responsibility of the member states (Art. 13 [EU07/715]), which had to report to the EU by January 2, 2009 on the measures introduced.

In Germany, the responsibility lies with the Federal Ministry of Transport and Digital Infrastructure as the supreme federal authority or the Federal Motor

Transport Authority (KBA) as the federal upper authority. Hints that the KBA received since 2014 were ignored, leading to public criticism of the KBA. The use of illegal defeat devices was not sufficiently pursued by German monitoring authorities because the introduction of effective sanctions required by the EU was omitted. Furthermore, the KBA granted type approval based on incomplete documents without ensuring that the legal requirements are met [Führ16]. It remains open whether these omissions were due to negligence or intent. The KBA's stubborn refusal to provide documents to clarify the facts could only be broken jurisdictionally [DUH21].

If non-compliance is not monitored and leads to adequate consequences, even the prescribed compliance with legal regulations is economically unattractive and is omitted by some manufacturers, as recent events have shown. The manufacturers and their associations strive to minimize legal requirements through political influence, see Sect. 6.4.

6.3   Integration into Transport and Environmental Policy

As engineers we work on technical improvements to the vehicle. These can make a significant contribution to air pollution control, but they do not replace a transport policy that ensures high mobility with minimal environmental impact. Even such a transport policy is only part of a comprehensive environmental policy, as the pollutants from transport represent only a section of the total of all pollutants.

Road traffic alone includes not only car traffic but also traffic with trucks, two-wheelers, and buses. While extensive measures to reduce pollutants have already been taken in car and truck traffic, there is a high need for action for motorized two-wheelers, which emit high amounts of pollutants compared to cars and trucks.

A special traffic policy measure in Germany are the traffic bans established since 2008 according to [BImSchV35], also known as environmental zones or badge zones. Whether a vehicle is exempt from such a traffic ban depends on whether the vehicle carries a green badge (in Neu-Ulm, the yellow badge has been sufficient for many years). Although model calculations predicted impressive emission reductions, the zones did not noticeably improve the measured air quality [Laberer09, Morfeld13, Faßbender17] or only to a small extent [Boltze14]. Many model calculations that predicted a positive effect on air quality were based on the *Handbook Emission Factors for Road Transport* [HBEFA]. This handbook (actually a database) assumed that deviations between traffic and test stand are below the now known extent. However, studies show higher deviations for almost all brands [BMVI16, T&E16] and refute the model calculations used. After PEMS measurements in version 3.3 and 4.1 of the HBEFA led to significant adverse shifts compared to old versions (version 4.2 from 2022 takes into account updates and thus improvements again), a repetition of the model calculations is necessary. In addition to various possible reasons why the zones do not work as hoped, it was already pointed out in [Borgeest11] that the classification of vehicles and thus the allocation of a badge is based on an unrealistic procedure (Sect. 3.1). Introducing a blue badge, which was originally planned and discarded in 2016, before the introduction of more realistic certification procedures, could have potentially harmed air quality in the long term. It would have caused many more dirty Euro-6 new registrations, which are only cleaner on paper and would then have remained in traffic for many years. Without a blue badge, a larger part of the uncontrolled, age-related vehicle fluctuation with subsequent new purchase now falls into the time of more realistic certification procedures from September 2017. In the meantime, municipalities are planning to abolish these zones [SWR22].

In addition to traffic bans, there are diesel transit restrictions for individual road sections or areas in several German cities.

A similar problem as with the badges also occurs with a pollutant-dependent car toll, if manipulated or unrealistically tested values are used instead of the actual emissions.

There are indications that the Federal Ministry of Transport has made gross errors in the environmental impact assessment of federal highways [Hahn22]. A Federal Transport Infrastructure Plan, which was unlawfully created in this way, can be replaced relatively quickly by a sensible, fact-based plan.

The increase in freight traffic is taken for granted, but is by no means a prerequisite for supporting economic growth. On the one hand, spatially widely distributed supply chains must be questioned, they not only burden the environment and the roads, they are also vulnerable, as shown for example by the reintroduction of border controls at EU internal borders in 2015 and the restrictions due to the Corona pandemic in 2020. Since 2021, 506 billion tkm of goods were transported on the road in Germany, but only 131 billion tkm on the railway and 48 billion tkm by inland waterway [destatis], it is necessary to shift long-distance freight traffic to the railway on a large scale, as Switzerland does particularly effectively, in individual cases a shift to inland waterways can also make sense. So far, no Federal Minister for Transport has been successful in this respect.

The acceptance of the railway in passenger transport also depends on whether trains run according to schedule as in other countries.

Regarding pollutant emissions and CO_2 emissions a very effective and at the same time simple and short-term implementable[1] measure is a speed limit. Depending on the modeling, a calculation of the effectiveness will come to different values. According to a study by the Federal Environment Ministry, a speed limit of e.g. 120 km/h not only leads to a CO_2 saving of almost 7 Mt/year but also to reductions in other pollutants [UBA23].

In addition to road traffic, there are also legal and technical efforts to reduce emissions from other modes of transport.

There are other sources of typical traffic emissions such as particles and nitrogen oxides. For example, the number of wood-burning heaters increased temporarily after 2000. Since these have a more favorable CO_2 balance than fossil fuels, these conversions are still partly promoted. This increases particle emissions, especially in residential areas, see also [UBA23P]. Thus, efforts to reduce vehicle particle emissions will have only a minor impact on immissions.

6.4   Influence and Corruption

In the previous chapters, the progress in technology and the legal processing of the emissions scandal were already described. The emissions scandal was made possible in the first place by democratically questionable entanglements between industry and politics. These entanglements are apparently robust enough to survive the emissions scandal unchanged so far. Unfortunately, the personnel and party change at the head of the Federal Ministry of Transport in 2021 did not lead to a clean new start, the new Minister of Transport Wissing is instead trying to conceal his appointments and agreements with Porsche [AW22].

In the EU, influence is officially exerted by the *accredited lobbyists* of the automotive industry in the *Transparency Register* of the EU Commission and the EU Parliament. The car lobbyists represent about 1% of the accredited lobbyists. How many lobbyists in Brussels in total, including non-accredited lobbyists, control our legislation is not known. A concentration of power on a rather small circle like the Commission increases the risk of external control. External control of politics covers a wide range from politically reprehensible lobbying over the

---

[1] According to Federal Minister of Transport Wissing, Germany would not have enough traffic signs for this. This argument is not convincing because a nationwide speed limit does not require local signage.

wide gray area of legalized corruption to illegal corruption. The European Commission is accused, for example, by organizations like Lobbycontrol of having pursued the interests of lobbyists in matters of emissions legislation [Bank15]. Among other things, to clarify these processes in connection with the emissions scandal, the EU Parliament's Committee of Inquiry also questioned the former Commissioners Stavros Dimas (Environment from 2004 to 2010) and Günter Verheugen (Industry from 2004 to 2010) [DimasEU, VerhgEU].

In Germany, the work of lobbyists is even less transparent. A transparency register like in Brussels failed for a long time due to political will, but was then introduced in 2021 under the name "Lobby Register" catalyzed by the mask affair. Since many forms of corruption by parties are legalized by the legislator, punishable corruption offenses are likely to play a rather minor role. Significant ways of influencing include, for example, party donations, exhibits and stands at party events, the sometimes high fees of which do not have to be published as party donations [AW17], further forms of party financing and personnel exchange between ministries and boards/supervisory boards. However, it is not always disadvantageous when experienced managers from industry move into areas close to politics.

6.5 Potential Improvements

The emissions scandal highlighted problems in the political process, especially:

- the influence on German and European legislative bodies,
- a failure of control mechanisms.

Parties are places of democratic decision-making, they occupy state organs and control them. The decision-making process must originate exclusively from the people, not from donors. This requires absolute transparency to the public about party finances. There should be no secrets or privacy here.

Influencing European legislative bodies is the professional task of lobbyists paid and active in Brussels. Strengthening the democratically legitimized representation of the people in the form of the European Parliament could bring the public interest more to the fore.

It is a difficult task to exclude failure of monitoring mechanisms in future, as the beneficiaries of the failure continue to exert an influence on constitutional bodies not provided for by the German constitution, which can only be countered by public pressure. The deputy chairman of the Bundestag's investigative committee coined the term "organized state failure" [Lindner16 and other sources] adopted by many media to express a possible intent. Ultimately, the Bundestag's investigative committee was also unable to clarify why the responsible Department 4 of the Federal Motor Transport Authority approved vehicles that are not eligible for approval under EU law and thus also German law [D18/12900]. The announcement by the Federal Ministry of the Environment to monitor compliance with emission values and thus indirectly the KBA sounds initially like the creation of redundant structures, since the KBA is subordinate to the Ministry of Transport [Hornung16]; in fact, effective monitoring of the KBA supervisory authority would probably have prevented the emissions scandal. It is worrying that a public prosecutor's office had to threaten the KBA with a search because it was withholding evidence in the emissions scandal [Magenh19]. The long-term goal should be to establish a sovereignly acting, uninfluenced supervisory authority in Germany; whether this is still the KBA in the area of responsibility of the Ministry of Transport or in the future in the area of responsibility of the Ministry of the Environment will certainly also depend on how independent and sovereign the two ministries themselves are. The American EPA can serve as a model, although the previous US President Donald Trump tried to weaken it.

In addition to these structural problems, the environmental policy task remains to reduce pollutants as a whole; beyond those vehicles which are still not sufficiently covered by the emissions legislation this includes also sources outside of transport.

References

[AW17]   abgeordnetenwatch.de, M Ruddat: *Wie viel Lobbyisten für einen Stand auf einem Parteitag zahlen*, 02.03.2017. https://www.abgeordnetenwatch.de/blog/lobbyismus/wie-viel-lobbyisten-fuer-einen-stand-auf-einem-parteitag-zahlen. Accessed: 21. Febr. 2023

[AW22]   abgeordnetenwatch.de, M. Reyher: *Wissing-Ministerium will Lobbyisten vor Transparenz bewahren*, 21.10.2022. https://www.abgeordnetenwatch.de/recherchen/lobbyismus/wissing-ministerium-will-lobbyisten-vor-transparenz-bewahren. Accessed: 21. Febr. 2023

[Bank15]   Bank, M.: *Die Macht der deutschen Autolobby in Brüssel*, article on Lobbycontrol homepage, 29.09.2015. https://www.lobbycontrol.de/2015/09/die-macht-der-deutschen-autolobby-in-bruessel. Accessed: 16. Febr. 2023

[BImSchV35] *Fünfunddreißigste Verordnung zur Durchführung des Bundes-Immissionsschutzgesetzes,* Verordnung zur Kennzeichnung der Kraftfahrzeuge mit geringem Beitrag zur Schadstoffbelastung vom 10.

Oktober 2006 (BGBl. I S. 2218), die zuletzt durch Artikel 85 der Verordnung vom 31. August 2015 (BGBl. I S. 1474) geändert worden ist. http://www.gesetze-im-internet.de/bimschv_35. Accessed: 15. Febr. 2023

[BMVI16]   Bundesministerium für Verkehr und digitale Infrastruktur: *Bericht der Untersuchungskommission „Volkswagen"* (2016). https://www.kba.de/DE/Themen/Marktueberwachung/Abgasthematik/erster_ber_uk_vw_nox.pdf; jsessionid=472871D156856711E618BD0BA98DE6D0.live11292?__blob=publicationFile&v=2. Accessed: 15. Febr. 2023

[Boltze14]   Boltze, M., Jiang, W., Groer, S., Scheuvens, D: Analyse der *Wirksamkeit von Umweltzonen hinsichtlich Feinstaub- und Stickstoffoxidkonzentrationen*, Straßenverkehrstechnik **4** (2014), 219–228 https://www.verkehr.tu-darmstadt.de/media/verkehr/fgvv/prof_boltze/BoVeroeff165.pdf. Accessed: 21. Febr. 2023

[Borgeest11]   Borgeest, K.: *Technische Zusammenhänge zur Bewertung rechtlicher Folgen der Einrichtung plakettenpflichtiger Verkehrszonen*, NVwZ 16 (2011), PDF: https://rsw.beck.de/rsw/upload/NVwZ/NVwZ-Extra_2011_16.pdf. Accessed: 21. Febr. 2023

[destatis]   Statistisches Bundesamt (Federal Statistical Office): Table. https://www.destatis.de/DE/Themen/Branchen-Unternehmen/Transport-Verkehr/Gueterverkehr/Tabellen/gueterbefoerderung-lr.html. Accessed: 16. Febr. 2023

[DimasEU]   Dimas, S.: Statement to committee of inquiry EMIS of the European Parliament (14.07.2016). http://www.europarl.europa.eu/committees/en/emis/events-hearings.html?id=20160714CHE00302. Accessed: 30. May 2023

[DUH21]   Deutsche Umwelthilfe: *Deutsche Umwelthilfe veröffentlicht brisante Dieselgate-Akten des Kraftfahrt-Bundesamts und des Bundesverkehrsministeriums*, Press Release, 23.04.2021. https://www.duh.de/presse/pressemitteilungen/pressemitteilung/deutsche-umwelthilfe-veroeffentlicht-brisante-dieselgate-akten-des-kraftfahrt-bundesamts-und-des-bund. Accessed: 21. Febr. 2023

[D18/12900]   *Beschlussempfehlung und Bericht des 5. Untersuchungsausschusses gemäß Artikel 44 des Grundgesetzes*, Circular 18/12900. https://dip21.bundestag.de/dip21/btd/18/129/1812900.pdf. Accessed: 21. Febr. 2023

[EU07]   *Treaty of Lisbon amending the Treaty on European Union and the Treaty establishing the European Community*, signed at Lisbon 13 December 2007. https://eur-lex.europa.eu/legal-content/EN/TXT/PDF/?uri=OJ:C:2007:306:FULL. Accessed: 21. Febr. 2023

[EU07/715]   *Regulation (EC) No 715/2007 of the European Parliament and of the Council of 20 June 2007 on type approval of motor vehicles with respect to emissions from light passenger and commercial vehicles (Euro 5 and Euro 6) and on access to vehicle repair and maintenance information.* https://eur-lex.europa.eu/legal-content/EN/TXT/PDF/?uri=CELEX:32007R0715. Accessed: 21. Febr. 2023

[EU19/631]   *Regulation (EU) 2019/631 of the European Parliament and the Council of 17 April 2019 setting CO_2 emission performance standards for new passenger cars and for new light commercial vehicles, and repealing Regulations (EC) No 443/2009 and (EU) No 510/2011.* Accessed: 21. Febr. 2023.

[EUV92]   *Treaty on the European Union*, Maastricht (1992), consolidated version 2012. https://eur-lex.europa.eu/legal-content/EN/TXT/HTML/?uri=CELEX:120 12E/TXT. Accessed: 21. Febr. 2023

[Faßbender17] Faßbender, K.: *Der Dieselskandal und der Gesundheitsschutz*, NJW, S. 1995 (2017)

[Führ16]   Führ, M.: *Gutachterliche Stellungnahme für den Deutschen Bundestag – 5. Untersuchungsausschuss der 18. Wahlperiode, fortgeschriebene Fassung*, 19.11.2016.   https://www.bundestag.de/resource/blob/481344/c6f582c85 98c9d6b62fcfb2acd012462/stellungnahme-prof--dr--fuehr--sv-4--data.pdf. Accessed: 15. Febr. 2023

[Hahn22]   Hahn, W., Hoppe, R., Schleicher, P.: *Kurzstudie über Klimaschutzbeiträge zur Umweltverträglichkeitsprüfung von Bundesfernstraßen im Rahmen der Aufstellung des Bundesverkehrswegeplans 2040*, RegioConsult, Marburg, https://www.bund-naturschutz.de/fileadmin/Bilder_und_Dokumente/Pre sse_und_Aktuelles/2023/Mobilität/2022-11_Kurzstudie_Klimaschutzbeitr äge_des_Verkehrs_im_BVWP.pdf. (21.02.2022)

[HBEFA]   Infras: Handbook Emission Factors for Road Transport. https://www.hbefa. net/e/index.html. Accessed: 23. Febr. 2023

[Hornung16]   Hornung, P., Riedel, K.: *Kampfansage vom Umweltbundesamt*, 26.06.2016. https://www.tagesschau.de/wirtschaft/umweltbundesamt-abgastests-101. html. Abruf 20.12.2016, nicht mehr verfügbar

[Laberer09]   Laberer, C., Niedermaier, M.: *Wirksamkeit von Umweltzonen*, Studie, ADAC (2009).

[Lindner16]   Lindner, N.: *Audi und Abgas-Affäre „Ganz ohne Bescheißen" geht's nicht*, Deutschlandfunk (2016). https://www.deutschlandfunk.de/audi-und-abgas-affaere-ganz-ohne-bescheissen-geht-s-nicht-100.html. Accessed: 21. Febr. 2023

[LRV-CH]   Schweiz, *Luftreinhalte-Verordnung (LRV) vom 16. Dezember 1985* (Stand am 1. Januar 2023). https://www.admin.ch/opc/de/classified-compilation/ 19850321/index.html. Accessed: 21. Febr. 2023

[Magenh19]   Magenheim, Th.: *Neue Enthüllungen: Hat das Kraftfahrt-Bundesamt die Diesel-Schummler bei Audi geschützt?* Redaktionsnetzwerk Deutschland. https://www.rnd.de/wirtschaft/neue-enthullungen-hat-das-kraftfahrt-bun desamt-die-diesel-schummler-bei-audi-geschutzt-FH2QYDYKJRUDGGA B3RNWR7VI6Q.html. Accessed: 21. Febr. 2023

[Morfeld13]   Morfeld, P., Stern, R., Builtjes, P., Groneberg, D.A., Spallek, M.: *Einrichtung einer Umweltzone und ihre Wirksamkeit auf die PM10-Feinstaubkonzentration – eine Pilotanalyse am Beispiel München*. Zbl Arbeitsmed **63** (2013), 104–115

[StVO]   *Straßenverkehrs-Ordnung vom 6. März 2013 (BGBl. I S. 367), die zuletzt durch Artikel 13 des Gesetzes vom 12. Juli 2021 (BGBl. I S. 3091) geändert worden ist.* https://www.gesetze-im-internet.de/stvo_2013. Accessed: 21. Febr. 2023

[StVZO]   *Straßenverkehrs-Zulassungs-Ordnung vom 26. April 2012 (BGBl. I S. 679), die zuletzt durch Artikel 11 des Gesetzes vom 12. Juli 2021 (BGBl. I S.*

3091) geändert worden ist. https://www.gesetze-im-internet.de/stvzo_2012. Accessed: 21. Febr. 2023

[SWR22]   SWR: *In Baden-Württemberg sollen Umweltzonen wegfallen – weil die Luft besser geworden ist.* 23.12.2022. https://www.swr.de/swraktuell/baden-wue rttemberg/umweltzonen-in-bw-sollen-wegfallen-100.html. Accessed: 21. Febr. 2023

[T&E16]   Transport & Environment: *Dieselgate 1st anniversary: all diesel car brands in Europe are even more polluting than Volkswagen* – study, 19.09.2016. https://www.transportenvironment.org/press/dieselgate-1st-anniversary-all-diesel-car-brands-europe-are-even-more-polluting-volkswagen. Accessed: 21. Febr. 2023

[UBA23P] Umweltbundesamt: *Staub (PM2,5)-Emissionen nach Quellkategorien*, Chart (2023). https://www.umweltbundesamt.de/bild/staub-pm25-emissionen-nach-quellkategorien. Accessed: 16. June. 2023

[UBA23]   Umweltbundesamt (Hrsg.): *Flüssiger Verkehr für Klimaschutz und Luftreinhaltung*, Study (2023). https://www.umweltbundesamt.de/sites/default/files/medien/479/publikationen/texte_14-2023_fluessiger_verkehr_fuer_klimaschutz_und_luftreinhaltung.pdf. Accessed: 21. Febr. 2023

[VerhgEU] Verheugen, G.: Statement to committee of inquiry EMIS of the European Parliament (14.07.2016). https://www.europarl.europa.eu/committees/de/product-details/20160714CHE00302. Accessed: 21. Febr. 2023

The expected costs of non-compliance with legal regulations approach 0 when a manufacturer does not need to expect that violations will be detected and sanctioned. In the EU, many manufacturers have relied on violations not being detected and sanctioned, as they were long safe from serious regulatory oversight. Especially in Germany, there were no serious consequences for violations, as Germany had "forgotten" to define the consequences of violations required by [EU07/715] even after a reminder from Brussels. However, in 2018, the Braunschweig Public Prosecutor's Office issued a fine of one billion € against VW for negligent violation of supervisory duties, and other companies received fines of lesser amounts for the same reason (Sect. 5.6).

The economic consequences can be well observed in the case of Volkswagen, as many figures are now publicly known. According to [Cabraser16], Volkswagen wanted to aggressively conquer the American market with supposedly clean diesel engines, after Toyota's successful strategy with hybrid drives seemed too expensive. Unfortunately, this did not work without manipulations of the emission values for both the 2-l engines of the EA189 series sold in the USA and the 3-l engines. VW managers were naive to believe that American authorities would be as easy to deceive as the KBA. This may have led to the rift between Ferdinand Piëch and Martin Winterkorn [Focus16].

Since the measures presented in Chap. 4 to actually comply with emission limits are associated with costs, the use of defeat devices initially resulted in a cost advantage. Depending on which measure the engine would have required to comply with the limits in reality, development costs between 0 and millions of € per engine type were saved, as well as parts costs between 0 and an estimated

K. Borgeest, *Manipulation of Exhaust Gas Values*,
https://doi.org/10.1007/978-3-658-45864-5_7

2000 € per sold vehicle. It is difficult to quantify the saved costs for two reasons. On the one hand, knowledge of the manufacturer's tests on the engine test bench and roller test bench is required to be able to estimate with high reliability which measures would have been necessary on the specific engine to achieve the permissible emissions legally. The second difficulty is that there are no price lists for these measures. Supplier prices are negotiated on a case-by-case basis. Manufacturers keep relevant information confidential. If we take the additional costs for a complete SCR system as an example, these are stated to be between a few 100 € and a few 1000 €.

Manufacturers who have manipulated now face high costs due to penalties in the USA, due to retrofitting affected vehicles and further civil law claims from defrauded buyers, also mainly in the USA. Since the principle in the USA is that the penalties should not be less than the illegally achieved cost advantages, an economic disadvantage should also result here in the end. Indirectly, sales figures are falling. Demand for diesel cars in the USA, where diesel has traditionally been a niche product with a few % market share, completely collapsed, but the market share is also falling in Europe. In Germany, the market share fell from a peak of 48% (2015) to 32% (2019), and the development is similarly pronounced in other European countries [CCFA21].

On 25.10.2016, VW was able to limit the civil law costs for the manipulations on the EA189-2-l engines in the USA to 13.5 billion € (approx. 15 billion US$) through a settlement, of which 9.2 billion € for repurchase and compensation plus 4.3 billion € (environmental fund, development of emission-free cars) [SFrancisco16]. On 15.11.2016, there was a preliminary agreement in the USA on how the owners of 80,000 manipulated 3-l engines would be compensated, which was finalized in December. Not included in this settlement are the manipulated engines of the EA288 series [CARB15]. In addition, there are criminal costs of over 4 billion €. In Canada, VW agreed to a settlement of approx. 1.5 billion € for the 2-l engines on 20.12.2016 [VW16].

In Germany, after the model declaratory proceedings, about a quarter of a million buyers were compensated with approx. 750 million € [Hahne21], but there are still too many individual proceedings pending to quantify the total compensation. The expiry of the operating license for manipulated vehicles [Führ16] leads to significant compensation claims by the owners (Chap. 5) and thus also to high costs in Europe, but software updates were able to maintain the operating license and limit the damage. The compensations for the remaining loss in value have remained low in Germany; many owners have waived their right to return against compensation for use by not making claims or accepting a cheap settlement for VW. Since many claims for damages against VW are still pending, the economic damage cannot yet be assessed, but even in the worst case, the costs for VW, despite high unit numbers on the local market, are likely to remain below those in the USA. VW itself estimates the total damage caused by the emissions scandal at approx. 32 billion € [Hahne21].

Daimler settled in 2021 and paid 175 million to Canadian customers [Handelsbl21]. In the USA, Daimler paid 1.27 billion € for settlements with authorities and 592 million € for the settlement of a class action lawsuit by consumers, these expenses are not exhaustive [Baumann21].

However, in manipulations that are still within the legal gray area, the realization of an economic advantage is possible.

In addition to the direct costs, indirect losses in revenue due to image loss are to be expected. In particular, the once high reputation of "Made in Germany" was damaged in the USA, extending beyond the automotive industry.

The manipulations affect the stock price of the involved companies, violations of the obligations according to [AktG] justify compensation to the shareholders, which causes further costs; such claims for damages from VW's major shareholders, e.g. the Free State of Bavaria [BR16], are known; so far, there are no judgments on this.

There remains the hope that companies that survive the short-term consequences will see their involvement in the emissions scandal as a stimulus for a new beginning, which can bring economic advantages in the very long term (e.g. the current beginning, long-term realignment of the VW Group towards electromobility). It is regrettable that it took this scandal to provide the impetus for the realignment.

References

[AktG]   *Aktiengesetz vom 6. September 1965 (BGBl. I S. 1089), das zuletzt durch Artikel 3 Absatz 3 des Gesetzes vom 4. Januar 2023 (BGBl. 2023 I Nr. 10) geändert worden ist.* http://www.gesetze-im-internet.de/aktg. Accessed: 6. Febr. 2023

[Baumann21]   Baumann, U.: *US-Dieselvergleich kostet 1,9 Milliarden Euro.* https://www.auto-motor-und-sport.de/verkehr/daimler-abgasskandal-us-dieselvergleich-kostet-1-9-milliarden-euro. Accessed: 6. Febr. 2023

[BR16]   Bayerischer Rundfunk, Siller, M., Gamböck, B.: *Warum Bayern VW verklagt.* http://www.br.de/nachrichten/bayern-verklagt-vw-100.html. Accessed: 6. Febr. 2023

[Cabraser16] Cabraser, E.J.: Complaint. http://www.cand.uscourts.gov/filelibrary/1709/
Consolidated_Consumer_Class_Action_Complaint.pdf.   Accessed: 21.
Febr. 2023

[CARB15] CARB, Letter of 18.09 (2015). https://www.arb.ca.gov/newsrel/in_use_com
pliance_letter.htm. Accessed: 20. Dec. 2016, no longer available

[CCFA21] Comité des Constructeurs Français d'Automobiles: *Anteil der Dieselautos
an den Pkw-Neuzulassungen in ausgewählten Ländern Europas von 2012
bis 2019*, (2021) in Statista. https://de.statista.com/statistik/daten/studie/
30846/umfrage/dieselanteil-an-den-pkw-neuzulassungen-in-westeuropa.
Accessed: 6. Febr. 2023

[EU07/715] Regulation (EC) No 715/2007 of the European Parliament and of the
Council of 20 June 2007 on type approval of motor vehicles with respect to
emissions from light passenger and commercial vehicles (Euro 5 and Euro
6) and on access to vehicle repair and maintenance information. https://
eur-lex.europa.eu/legal-content/DE/TXT/PDF/?uri=CELEX:32007R0715.
Accessed: 6. Febr. 2023

[Focus16] „In Haus in Salzburg befragt, als Piëch Winterkorn auf den Abgas-Skandal
ansprach, ließ der ihn nur abblitzen", Focus (28.08.2016). https://www.
focus.de/auto/news/in-haus-in-salzburg-befragt-als-piech-winterkorn-auf-
den-abgas-skandal-ansprach-liess-der-ihn-nur-abblitzen_id_5868230.html.
Accessed: 6. Febr. 2023

[Führ16] Führ, M.: *Gutachterliche Stellungnahme für den Deutschen Bundestag – 5.
Untersuchungsausschuss der 18. Wahlperiode*, fortgeschriebene Fassung
(19.11.2016).   https://www.bundestag.de/resource/blob/481344/c6f582c85
98c9d6b62fcfb2acd012462/stellungnahme-prof--dr--fuehr--sv-4--data.pdf.
Accessed: 6. Febr. 2023

[Hahne21] Hahne, S.: *Die schleppende Aufarbeitung des VW-Diesel-Skandals*. Deutschlandfunk   (19.04.2021).   https://www.deutschlandfunk.de/dieselgate-und-
die-folgen-die-schleppende-aufarbeitung-des.724.de.html?dram:article_
id=495928. Accessed: 6. Febr. 2023

[Handelsbl21] Handelsblatt: *Daimler schließt in Kanada Vergleich mit Diesel-Klägern*,
10.12.2021,   https://www.handelsblatt.com/unternehmen/industrie/sam
melklage-daimler-schliesst-in-kanada-vergleich-mit-diesel-klaegern/278
83988.html. Accessed: 6. Febr. 2023

[SFrancisco16] U.S. District Court: 15-md-02672-CRB. Northern District (San Francisco),
(2016)

[VW16] Volkswagen Group: https://www.vwcanadasettlement.ca/en.   Accessed
20.12.2016, no longer available